T0183406

# Lecture Notes in Computer Science 8807

Commenced Publication in 1973
Founding and Former Series Editors:
Gerhard Goos, Juris Hartmanis, and Jan van Leeuwen

## Editorial Board

David Hutchison
  *Lancaster University, Lancaster, UK*
Takeo Kanade
  *Carnegie Mellon University, Pittsburgh, PA, USA*
Josef Kittler
  *University of Surrey, Guildford, UK*
Jon M. Kleinberg
  *Cornell University, Ithaca, NY, USA*
Friedemann Mattern
  *ETH Zurich, Zürich, Switzerland*
John C. Mitchell
  *Stanford University, Stanford, CA, USA*
Moni Naor
  *Weizmann Institute of Science, Rehovot, Israel*
C. Pandu Rangan
  *Indian Institute of Technology, Madras, India*
Bernhard Steffen
  *TU Dortmund University, Dortmund, Germany*
Demetri Terzopoulos
  *University of California, Los Angeles, CA, USA*
Doug Tygar
  *University of California, Berkeley, CA, USA*
Gerhard Weikum
  *Max Planck Institute for Informatics, Saarbruecken, Germany*

More information about this series at http://www.springer.com/series/7409

Jianfeng Zhan · Rui Han
Chuliang Weng (Eds.)

# Big Data Benchmarks, Performance Optimization, and Emerging Hardware

4th and 5th Workshops, BPOE 2014
Salt Lake City, USA, March 1, 2014
and Hangzhou, China, September 5, 2014
Revised Selected Papers

 Springer

*Editors*
Jianfeng Zhan
ICT, Chinese Academy of Sciences
Beijing
China

Chuliang Weng
Shannon (IT) Lab.
Huawei
China

Rui Han
ICT, Chinese Academy of Sciences
Beijing
China

ISSN 0302-9743          ISSN 1611-3349  (electronic)
Lecture Notes in Computer Science
ISBN 978-3-319-13020-0     ISBN 978-3-319-13021-7  (eBook)
DOI 10.1007/978-3-319-13021-7

Library of Congress Control Number: 2014953862

Springer Cham Heidelberg New York Dordrecht London

© Springer International Publishing Switzerland 2014
This work is subject to copyright. All rights are reserved by the Publisher, whether the whole or part of the material is concerned, specifically the rights of translation, reprinting, reuse of illustrations, recitation, broadcasting, reproduction on microfilms or in any other physical way, and transmission or information storage and retrieval, electronic adaptation, computer software, or by similar or dissimilar methodology now known or hereafter developed.
The use of general descriptive names, registered names, trademarks, service marks, etc. in this publication does not imply, even in the absence of a specific statement, that such names are exempt from the relevant protective laws and regulations and therefore free for general use.
The publisher, the authors and the editors are safe to assume that the advice and information in this book are believed to be true and accurate at the date of publication. Neither the publisher nor the authors or the editors give a warranty, express or implied, with respect to the material contained herein or for any errors or omissions that may have been made.

Printed on acid-free paper

[Springer International Publishing AG Switzerland] is part of Springer Science+Business Media
(www.springer.com)

# Preface

Today, huge amounts of data are being collected in many areas, which create new opportunities to understand phenomena in meteorology, health, finance, and many other sectors. Big Data is considered precious assets of companies, organizations, and even nations. Turning such big data into real treasures requires the support of big data systems and platforms. However, the sheer volume of big data requires significant storage capacity, transmission bandwidth, computation, and power consumption. It is expected that systems with unprecedented scales can resolve the problems caused by varieties of big data with daunting volumes.

The complexity, diversity, frequently changed workloads, and rapid evolution of big data systems raise great challenges in big data benchmarking. Without big data benchmarks, it is very difficult for big data owners to make a decision on which system is best for meeting with their specific requirements. They also face challenges on how to optimize the systems and their solutions for specific or even comprehensive workloads. Meanwhile, researchers are also working on innovative data management systems, hardware architectures, operating systems, and programming systems to improve performance in dealing with big data.

This book includes papers from two workshops, which are the fourth and fifth workshops on Big Data Benchmarks, Performance Optimization, and Emerging Hardware (BPOE-4 and BPOE-5). BPOE-4 (http://prof.ict.ac.cn/bpoe_4_asplos/) is co-located with ASPLOS 2014 (http://www.cs.utah.edu/asplos14/), a premier conference on architecture support for operating systems and programming systems. BPOE-5 (http://prof.ict.ac.cn/bpoe_5_vldb/) is co-located with VLDB 2014 (http://www.vldb.org/2014/), a premier conference on data management, database and information systems. Both workshops focus on architecture and system support for big data systems, aiming at bringing researchers and practitioners from data management, architecture, and systems research communities together to discuss the research issues at the intersection of these areas.

The call for papers for these two workshops attracted a number of high-quality international submissions. Within a rigorous process, in which each paper was reviewed by at least four experts, we selected 6 papers out of 12 submissions for inclusion in the BPOE-04 and 10 papers out of 18 submissions in the BPOE-05, respectively. In addition, several prestigious keynote speakers were invited, including Prof. Lizy Kurian John at University of Texas at Austin (http://users.ece.utexas.edu/~ljohn/) whose topic was "Big Data Workloads: An Architect's Perspective," Prof. Dhabaleswar K. (DK) Panda at Ohio State University (http://www.cse.ohio-state.edu/~panda/) whose topic was "Accelerating Big Data Processing with RDMA-Enhanced Apache Hadoop," Prof. Christos Kozyrakis at Stanford University (http://csl.stanford.edu/~christos/) whose topic was "Resource Efficient Cloud Computing," and Dr. Jeff Stuecheli from IBM (http://www.linkedin.com/pub/jeff-stuecheli/2/664/a0a) whose topic was "Power Technology For a Smarter Future."

We are very grateful to the efforts of all authors related to writing, revising, and presenting their papers at BPOE workshops. Finally, we appreciate the indispensable support of BPOE Program Committees and thank their efforts and contributions in maintaining the high standards of the BPOE workshop.

August 2014                                                            Jianfeng Zhan
                                                                             Rui Han
                                                                        Chuliang Weng

# Organization

## Program Co-chairs

Jianfeng Zhan      ICT, Chinese Academy of Sciences and University of Chinese Academy of Sciences, China
Chuliang Weng      Shannon (IT) Lab, China
Rui Han      ICT, Chinese Academy of Sciences, China

## Steering Committee

Christos Kozyrakis      Stanford University, USA
Xiaofang Zhou      University of Queensland, Australia
Dhabaleswar K. (DK) Panda      Ohio State University, USA
Aoying Zhou      East China Normal University, China
Raghunath Nambiar      Cisco, USA
Lizy Kurian John      University of Texas at Austin, USA
Xiaoyong Du      Renmin University of China, China
Ippokratis Pandis      IBM Almaden Research Center, USA
Xueqi Cheng      ICT, Chinese Academy of Sciences, China
Bill Jia      Facebook, USA
Lidong Zhou      Microsoft Research Asia, China
H. Peter Hofstee      IBM Austin Research Laboratory, USA
Alexandros Labrinidis      University of Pittsburgh, USA
Cheng-Zhong Xu      Wayne State University, USA
Guang R. Gao      University of Delaware, USA
Yunquan Zhang      ICT, Chinese Academy of Sciences, China

## Program Committee

Onur Mutlu      Carnegie Mellon University, USA
Xu Liu      Rice University, USA
Meikel Poess      Oracle Corporation, USA
Dejun Jiang      ICT, Chinese Academy of Sciences, China
Yueguo Chen      Renmin University, China
Rene Mueller      IBM, Almaden Research Center, USA
Xiaoyi Lu      Ohio State University, USA
Yongqiang He      Dropbox, USA
Edwin Sha      University of Texas at Dallas, USA
Kun Wang      IBM Research China, China
Rong Chen      Shanghai Jiao Tong University, China

| | |
|---|---|
| Jens Teubner | TU Dortmund University, Germany |
| Yinliang Yue | ICT, Chinese Academy of Sciences, China |
| Mauricio Breternitz | AMD Research, China |
| Seetharami Seelam | IBM, USA |
| Zhenyu Guo | MSRA |
| Farhan Tauheed | EPFL, Switzerland |
| Gansha Wu | Intel, China |
| Bingsheng He | Nanyang Technological University, Singapore |
| Zhibin Yu | SIAT, Chinese Academy of Sciences, China |
| Lei Wang | ICT, Chinese Academy of Sciences, China |
| Yuanchun Zhou | CNIC, Chinese Academy of Sciences, China |
| Tilmann Rabl | University of Toronto, Canada |
| Weijia Xu | TACC, University of Texas at Austin, USA |
| Mingyu Chen | ICT, Chinese Academy of Sciences, China |
| Jian Ouyang | Baidu, China |
| Wentao Qu | Google, USA |
| Guangyan Zhang | Tsinghua University, China |
| Cheqing Jin | East China Normal University, China |
| Jiuyang Tang | National University of Defense Technology, China |
| Farhan Tauheed | EPFL, Switzerland |
| Xiaoyu Zhang | CSHUST, USA |
| Lijie Wen | School of Software, Tsinghua University, China |
| Rong Chen | Shanghai Jiao Tong University, China |

# Contents

**Topical Section Headings: Emerging Hardware**

# Topical Section Headings:
# Benchmarking

# On Big Data Benchmarking

Rui Han[1]([✉]), Xiaoyi Lu[2], and Jiangtao Xu[3]

[1] Department of Computing, Imperial College London, London, UK
r.han10@imperial.ac.uk
[2] Ohio State University, Columbus, USA
luxi@cse.ohio-state.edu
[3] Beijing Jiaotong University, Beijing, China
11301168@bjtu.edu.cn

**Abstract.** Big data systems address the challenges of capturing, storing, managing, analyzing, and visualizing big data. Within this context, developing benchmarks to evaluate and compare big data systems has become an active topic for both research and industry communities. To date, most of the state-of-the-art big data benchmarks are designed for specific types of systems. Based on our experience, however, we argue that considering the complexity, diversity, and rapid evolution of big data systems, for the sake of fairness, big data benchmarks must include diversity of data and workloads. Given this motivation, in this paper, we first propose the key requirements and challenges in developing big data benchmarks from the perspectives of generating data with 4 V properties (i.e. volume, velocity, variety and veracity) of big data, as well as generating tests with comprehensive workloads for big data systems. We then present the methodology on big data benchmarking designed to address these challenges. Next, the state-of-the-art are summarized and compared, following by our vision for future research directions.

**Keywords:** Big data systems · Benchmark · Data · Tests

## 1 Introduction

Big data systems have gained unquestionable success in recent years and will continue its rapid development over the next decade. These systems cover many industrial and public service areas such as search engine, social network and e-commerce, as well as a variety of scientific research areas such as bioinformatics, environment, meteorology, and complex simulations of physics. Conceptually, big data are characterized by very large data volumes and velocities, diversity and variety (various data types, and complex data processing requirements). In the era of big data, these data require a new generation of big systems to capture, store, search, and analyze them within an acceptable elapsed time. The complexity, diversity, and rapid evolution of big data systems give rise to new challenges in how to compare their performance, energy efficiency, and cost effectiveness.

© Springer International Publishing Switzerland 2014
J. Zhan et al. (Eds.): BPOE 2014, LNCS 8807, pp. 3–18, 2014.
DOI: 10.1007/978-3-319-13021-7_1

Conceptually, a big data benchmark aims to generate *application-specific* workloads and tests capable of processing big data sets to produce meaningful evaluation results [18]. Considering the diversity of big data systems (e.g. there are three mainstream application domains, namely search engine, social network, and e-commerce, of internet service workloads), and the emergence of new systems driven by the exploration of big data value, covering diversity of workloads is the prerequisite to perform successful and efficient benchmarking tests. Within this context, we propose our insights into the requirements and challenges in developing big data benchmarks. We also present the methodology on big data benchmarking, which represents our thinking about how to address these challenges. Finally, we discuss state-of-the-art benchmarking techniques for big data systems and propose some future research directions. The aim of this paper, therefore, is to provide a foundation towards building a successful big data benchmark, and to stimulate productive thinking, investigation, and development in this research area.

## 2    Requirements and Challenges

Big data benchmarks are developed to evaluate and compare the performance of big data systems and architectures. Figure 1 shows the benchmarking process for big data systems that consists of five steps. At the *Planning* step, the benchmarking object, application domain, and evaluation metrics are determined. In the following two steps, the data and test used in evaluation are generated. Next, the benchmark test is conducted at the *Execution* step and the evaluation result is reported. At the last step, the benchmarking result is analysed and evaluated.

Successful and efficient benchmarking can provide realistic and accurate measuring of big data systems and thereby addressing two objectives. (1) Promoting the development of big data technology, i.e. developing new architectures (processors, memory systems, and network systems), innovative theories, algorithms, techniques, and software stacks to manage big data and extract their value and hidden knowledge. (2) Assisting system owners to make decisions for planning system features, tuning system configurations, validating deployment strategies, and conducting other efforts to improve systems. For example, benchmarking results can identify the performance bottlenecks in big data systems, thus optimizing system configuration and resource allocation. In this section, we present the requirements and challenges in building big data benchmarks.

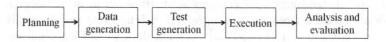

**Fig. 1.** Benchmarking process for big data systems

## 2.1   Generating Data with the 4V Properties of Big Data

Technically, a big data system can be attributed four dimensions: volume, velocity, variety, and veracity, which form the 4 V properties of big data [3]. (1) Volume represents the amount/size of data such as Terabyte (TB) or Petabyte (PB). (2) Velocity reflects the speed of generating, updating, or processing data. (3) Variety denotes the range of data types and sources. Finally, (4) veracity reflects whether the data used in benchmarking conform to the inherent and important characteristics of raw data.

In a big data benchmark, applying real-world data or generating synthetic data for application-specific workloads is a central problem. Traditionally, although some benchmarks use real data sets as inputs of their workloads and thereby guaranteeing the data veracity, the volume and velocity of real data sets cannot be flexibly adapted to different benchmarking requirements. Based on our experience, we also noticed that in many practical scenarios, obtaining a variety of real data is not trivial because many data owners are not willing to share their data due to confidential issues. In big data benchmarks, therefore, the consensus is to generate synthetic data as inputs of workloads on the basis of real data sets. Hence in synthetic data generation, preserving the 4 V properties of big data is the foundation of producing meaningful and credible evaluation results.

**Volume.** Today, data are generated faster than ever. For example, about 2.5 quintillion bytes of data are created every day [3] and this speed is expected to increase exponentially over the next decade according to International Data Corporation (IDC). In Facebook, there are 350 million photos updated and more than 500 TB data generated per day. The above facts indicate the big data generators must be able to generate different volumes of data as inputs of typical workloads. The data volume has different meanings in different workloads. For example, in workloads for processing text data (e.g. *sort* or *WordCount*), the volume is represented by the amount of data (e.g. 1 TB or 1 PB text data). In social network graph workloads, the volume is represented by the number of vertices in social graph data (e.g. $2^{20}$ vertices).

**Velocity.** In the context of big data generation, data velocity has threefold meanings. First of all, it represents the *data generation rate*. For example, generating 100 TB text data in 10 h means the generation rate is 10 TB/h. Secondly, many big data applications have real-time data updating; that is, data velocity represents the data *updating frequency* in this case. For example, in a social network site, the social graph data are continuously updating. Finally, in streaming processing systems, data streams continuously arrives and these streams must be processed in real-time to keep up with their arriving speed. Hence data velocity represents the processing speed. Given the above facts, it is challenging to reflect data generation rates, updating frequencies, and processing speeds in data generation.

**Variety.** The fast development of big data systems gives birth to a diversity of data types, which cover structured data (e.g. tables), unstructured data (e.g. text, graph, images, audios, and videos), and semi-structured data (e.g. web logs, reviews, and resumes, where reviews and resumes contains both text and graph data). Hence in big data benchmarking, it is required that the data generators can support the whole spectrum of data types including structured, semi-structured, and unstructured data, as well as representative data sources such as table, text, stream, and graph. It is also required that these data generators can support the complexity and diversity of workloads and keep in pace with their frequent changes.

**Veracity.** Preserving data veracity is probably the most difficult job in synthetic data generation. In big data generators, it is challenging to scale up or down a synthetic data set while keeping this data similar to the real data [18]; that is, the important characteristics of raw data must be preserved in synthetic data generation. Data veracity is important to guarantee the reality and credibility of benchmarking results.

## 2.2    Generating Benchmarking Tests with Comprehensive Workloads

Margo Seltzer et al. pointed out that a testing result is meaningful only when applying an application-specific benchmarking test [18]. Also, the diversity and rapid evaluation of big data systems means it is challenging to develop big data benchmarks to reflect various workload cases. Hence in big data benchmarks, identifying the typical workload behaviours for an application domain is the prerequisite of evaluating big data systems. Furthermore, big data benchmarks must consider the diversity of workloads to cover different types of application domains, as well as automatically generate tests based on these workloads. We now discuss the key challenges in generating workloads and tests to evaluate big data systems from the *functional view* and the *system view*.

**Functional view.** Given the complexity and diversity of workload behaviours in current big data systems, it is reasonable to say that no single set of behaviors is representative for all applications. Hence, it is necessary to abstract from the behaviors of different workloads to a general approach. This approach should identify typical workload behaviours in representative application domains. From the functional view, these behaviors represent the system-independent outcome of processing data, thus allowing the comparison of systems of different types, e.g. a DBMS and a MapReduce system.

There are two challenges in developing this abstraction approach. First, the *operations* to process big data in a specific application domain need to be abstracted and their functions need to be identified. For example, *select*, *put*, *get*, and *delete* are abstracted operations in database systems to operate table data. Secondly, given a set of abstracted operations, *workload patterns* need to be abstracted to describe complex processing tasks by combining abstracted operations. One abstracted workload pattern can contain one or multiple abstract operations as well as their workflow. For example, an abstract pattern of a *SQL query* can contain *select* and *put* operations, in which the *select* operation executes first.

**System view.** The abstracted operations and patterns are designed to capture the system-independent user behaviours of workloads, i.e. the data processing operations and their sequences. Thus, an abstracted benchmark test can be constructed based on abstracted operations and patterns, and this test is independent of underlying systems and software stacks. From the system view, this abstract test can be implemented over different systems and thereby allows the comparison of systems of the same type. For example, an abstract test consisting of a sequence of *read, write,* and *update* operations can be used to compare different DBMSs.

## 2.3   Execution

To perform a fair, efficient, and successful benchmarking test, there are several requirements and challenges to be addressed at the *Execution* step.

**Adapting to different data formats.** Since the same type of data can be stored in multiple formats, e.g. texts can be stored in a simple text file or more complex formats as web pages and pdf, Big data benchmarks need to provide format conversion, which can transfer a data set into an appropriate format capable of being used as the input of a test running on a specific system.

**Portability to representative software stacks.** A software stack consists of a set of programs working together to provide a fully functional solution. Big applications and systems belonging to one application domain are built on the basis of one or multiple software stacks. Hence covering a broad spectrum of representative software stacks in benchmark tests, as well as avoiding being too costly or difficult to port a test to a specific software stack is another important issue we need to consider.

**Fair measurement.** A fair and sensible evaluation has twofold meanings. First, big data systems usually have many optional configurations, while these configurations have different combinations for optimal performance when the systems run in different hardware platforms. Hence using default configurations in measurement cannot guarantee fair measurement. That is to say, when comparing different big data systems in heterogeneous platforms, each system must be configured separately for fair comparison. For example, a big data system may have some specific configuration to improve its performance, thus such configuration is not suitable for fair measurement. Second, repeatability is another important requirement of the evaluation. This requirement means the parameters of hardware and software configurations must be stately clearly so that the same result can be obtained when the evaluation is repeated several times. In particular, in cloud environment, there are multiple virtual machines (VMs) running in one physical machine (PM) and competing for compute resources. Hence we need to develop a comprehensive evaluation mechanism to effectively identify and estimate the impact of resource competition on benchmarking results, and to avoid the uncertainties incurred by this resource competition.

**Extensibility.** The fast evolution of big data systems requires big data benchmarks not only keeping in pace with state-of-the-art techniques and underlying systems, but also taking their future changes into consideration. That is, big data benchmarks should be able to add new workloads or data sets with little or no change to the underlying algorithms and functions [19].

**Usability.** Usability reflects users experiences in using benchmarks and it is a combination of factors. In big data benchmarks, these factors include ease of deploying, configuring, and use; high benchmarking efficiencies; simple and understandable performance metrics; convenient user interfaces; and so on.

# 3   On Benchmarking Methodology

## 3.1   Layer Design of Big Data Benchmarks

Figure 2 shows the layered design of a big data benchmark with three layers: The *User Interface Layer* provides interfaces to assist system owners to specify their benchmarking requirements, such as the selected data, workloads, metrics and the preferred data volume and velocity.

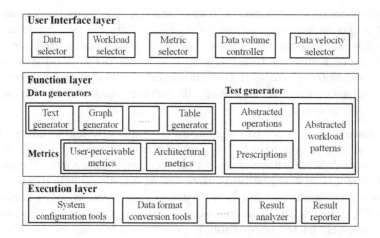

**Fig. 2.** Layered architecture of big data benchmarks

The *Function Layer* has three components: data generators, test generators and metrics. Briefly, data generators are designed to produce data sets covering different data types and application domains while keeping the 4 V properties of big data in these data sets. The test generator enables the automatic generation of tests with comprehensive workloads for big data systems. Metrics (either single or multiple metrics) can be divided into two types: user-perceivable metrics and architecture metrics [19]. User-perceivable metrics represent the metrics that matter for users; these metrics are usually observable and easy to be understood by users. Examples of user-perceivable metrics are the duration of a test, request latency, and throughput. While user-perceivable metrics are used to compare performances of workloads of the same category, architecture metrics are designed to compare workloads from different categories. Examples of architecture metrics are million instructions per second (MIPS) and million floating-point operations per second (MFLOPS). In addition, these metrics should not only measure system performance, but also take energy consumption, cost efficiency into consideration.

The *Execution Layer* offers several functions to support the execution of benchmark tests over different software stacks. Specifically, the system configuration tools enable a generated test running in a specify software stack. The data format conversion tools transform a generated data set into a format capable of being used by this test. The result analyzer and reporter display evaluation results.

## 3.2  Data Generators in Big Data Benchmarks

Data generators in big data benchmarks aim to efficiently generate data sets while preserving the 4 V properties of big data [15]. Figure 3 shows the process of generating data sets.

At the first step, data generators support the *variety* of big data by selecting real data sets to cover representative application domains as well as different data sources and types. The generators can also apply tools to directly generate synthetic data sets; that is, these synthetic data sets are independent of real data. This is because it is accepted that such purely synthetic data can be used as inputs of some workloads such as the *Sort* and *WordCount* workloads in Micro benchmarks; and the *Read*, *Write*, and Scan workloads belonging to basic database operations.

At the second step, each data generator employs a data model to capture and preserve the important characteristics in one or multiple real data sets of a specific date type. For example, a text generator can apply Latent dirichlet allocation (LDA) [9] to describe the topic and word distributions in text data. This generator first learns from a real text data set to obtain a word dictionary. It then trains the parameters and of a LDA model using this data set. Finally, it generates synthetic text data using the trained LDA model. To preserve data veracity, it is required that different models should be developed to capture the characteristic of real data of different types such as table, text, stream, and graph data. In addition, the sampling tools enable the scaling down of data set sizes.

At the third step, the volume and velocity can be controlled according to user requirements. For example, the data generation can be paralleled and distributed to multiple machines, thus supporting different data generation rates.

At the fourth step, after a data set is generated, the format conversion tools transform this data set into a format capable of being used as the input data of a specific workload.

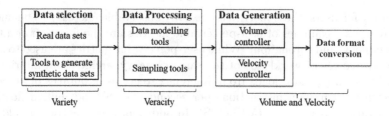

**Fig. 3.** The big data generation process

## 3.3   Test Generator in Big Data Benchmarks

In big data benchmarks, the test generator is developed to automatically generate benchmarking tests for big data systems. The basic idea of this generator is to abstract from the workload behaviours of current big data systems to a set of *operations* and *workload patterns* used in big data processing [20]. As shown in Fig. 2, the test generator consists of three components.

**Operations** represent the abstracted processing actions (operators) on data sets. In the test generator, we divide operations into three categories according to the number of data sets processed by these operations: element operation, single-set operation, and double-set operation.

**Workload patterns** are designed to combine operations to form complex processing tasks. In the test generator, we abstract three workload patterns: (1) a single-operation pattern contains one single operation; (2) a multi-operation pattern; and, (3) an iterative-operation pattern. The difference between a multi-operation pattern and an iterative-operation pattern is that the former pattern contains finite number of operations, while the latter pattern only provides stopping conditions that the exact number of operations can be known at run time.

**A prescription** includes the information needed to produce a benchmarking test, including data sets, a set of operations and workload patterns, a method to generate workload, and the evaluation metrics.

Figure 4 shows the process of generating a test. At steps 1, 2, and 3, a data set, a set of abstracted operations, and a set of workload patterns are selected, respectively. A prescription is then generated at step 4. Finally, at step 5, a prescribed test for a specific system and software stack is created based on the prescription and system configuration tools. Using the test generator, the workloads in different application domains can be automatically generated.

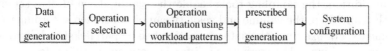

**Fig. 4.** The benchmark test generation process

# 4   State-of-the-Art

In this section, we review related work on big data benchmarks from the perspectives of data generation and benchmarking techniques.

## 4.1   Data Generation Techniques

As shown in Table 1, we now review data generation techniques in existing big data benchmarks according to the 4 V properties of big data.

**Volume.** To date, most of existing benchmarks generate synthetic data as their workload inputs, where the volume of synthetic data is scalable. By contrast, some benchmarks such as Hibench and LinkBench also use fixed-size data as inputs. Hence we call these benchmarks partially scalable in terms of data volume.

**Velocity.** Some benchmarks such as BigBench, LinkBench, and BigDataBench provide parallel strategies to support the deployment of multiple data generators. In these benchmarks, the data generation rate can be controlled. However, the equally important aspect of data velocity, the data updating frequency, is not considered in these benchmarks. Hence we call these benchmarks semi-controllable in terms of data velocity. We also called benchmarks un-controllable if both the data generation rate and updating frequency are not considered.

**Variety.** Table 1 lists the data sources of each benchmarks tested data, including tables (structured data); text, graph, and videos (unstructured data); and web logs and resumes (semi-structured data). We can observe that many current benchmarks only consider limited data types (e.g. the text data in Hibench or the table data in YCSB and TPC-DS). Although BigBench and CloudSuite benchmarks support a variety of data sources and types, they are only designed to test applications running in cloud service architecture, and DBMSs and MapReduce Hadoop, respectively.

**Veracity.** In GridMix, PigMix, YCSB, and Micro benchmark, the generation process of synthetic data is independent of the benchmarking applications. For example, in HiBench [13], the synthetic data sets are either randomly generated using the programs in the Hadoop distribution or created using some statistic distributions. Data veracity is un-considered in these benchmarks.

In TPC-DS, BigBench, LinkBench, and CloudSuite, the data generation tools partially consider the data veracity. For example, TPC-DS [12] implements a multi-dimensional data generator (MUDD). MUDD generates most of data using traditional synthetic distributions such as a Gaussian distribution. On the other hand, MUDD generates a small portion of crucial data sets using more realistic distributions derived from real data. In BigBench [12], table data are generated using PDGF [17], while web logs and reviews are generated on the basis of the table data. Hence the veracity of web logs and reviews rely on the table data.

By contrast, in BigDataBench [19], different data models are employ to capture and preserve the important characteristics of real data of different types (e.g. table, text, and table). The synthetic data are then generated using the constructed data model, thus avoiding the loss of data veracity.

In conclusion, the issues relating to keeping the 4 V properties of big data have not been adequately addressed by current big data benchmarks [1,2,4–6,8,10–14,16].

**Table 1.** Comparison of data generation techniques in existing big data benchmarks.

| Benchmark efforts | Volume | Velocity | Variety(data sources) | Veracity |
|---|---|---|---|---|
| Hibench [13] | Partially scalable | Un-controllable | Texts | Un-considered |
| GridMix [2] | Scalable | Un-controllable | Texts | Un-considered |
| PigMix | Scalable | Un-controllable | Texts | Un-considered |
| YCSB [10] | Scalable | Un-controllable | Tables | Un-considered |
| SIGMOD benchmark [16] | Scalable | Un-controllable | Tables, texts | Un-considered |
| TPC-DS [12] | Scalable | Semi-controllable | Tables | Partially Considered |
| BigBench [12] | Scalable | Semi-controllable | Texts, web logs, tables | Partially Considered |
| LinkBench [8] | Partially scalable | Semi-controllable | Graphs | Partially Considered |
| CloudSuite [11] | Partially scalable | Semi-controllable | Texts, resumes, graphs, tables | Partially Considered |
| BigDataBench [19] | Scalable | Semi-controllable | Texts, resumes, graphs, tables | Considered |

## 4.2 Benchmarking Techniques

Most of existing big data benchmarks aims to evaluate specific type of systems or architectures. As listed in Table 2, many benchmarks are developed to test the performance of DBMSs and Hadoop MapReduce, or compare the performance of both types of systems. Specifically, HiBench [13], GridMix [2] and PigMix [4] are designed to test MapReduce Hadoop systems. The SIGMOD benchmark in [16] compare two parallel SQL DBMSs (i.e. DBMS-X and Vertica) with MapReduce systems. TPC-DS is TPCs latest decision support benchmark [7] designed to test the performance of DBMSs in decision support systems. Adopting from TPC-DS by adding a web log generator and a review generator, BigBench aims to test the Teradata Aster DBMS and MapReduce Hadoop systems [12]. In [1], the benchmark is designed to test four SQL driven systems for managing data, including one database (Redshift), one data warehousing systems (Hive), and two engines (Spark and Impala). LinkBench tests MySQL databases that store Facebooks social graph data, and characterizes the real-world database workloads for social applications [8].

Some other benchmarks target at evaluating NoSQL databases or architectures. Yahoo! Cloud Serving Benchmark (YCSB) benchmark compares two non-relational databases (Cassandra and HBase) against one geographically distributed database (PNUTS) and a traditional relational database (MySQL) [10]. The CloudSuite benchmarks in [10; 13] are implemented to test cloud service architectures. Standard Performance Evaluation Corporation (SPEC) [6] has produced several benchmarks for evaluating workstations and has several server and client benchmarks including a Java business benchmark called SPECjBB, but specific big data benchmarks are not available. SPEC had produced web server benchmarks called SPECweb96, SPECweb99, SPECweb2005, and SPECweb2009, but they have been retired. The SPECjEnterprise 2010 and SPEC jBB benchmarks are the closest to big data/cloud benchmarks in SPECs suites.

At present, BigDataBench is the only big data benchmark that supports the evaluation of a hybrid of different big data systems. The workloads in BigData-Bench cover three fundamental and widely usage scenarios (i.e. micro benchmarks, "Cloud OLTP" workloads, and relational queries workloads) and three major application domains in internet services (i.e. Search Engine, Social Network, and E-commerce).

From the perspective of applications users, Table 2 divides workloads in current big data benchmarks into three categories. (1) Online services: these services are sensitive to the response delay, i.e. the time interval between the arrival and departure moments of a service request. Examples of workloads belonging to this category are typical MapReduce operations such as *sort* and *WordCount*. (2) Offline analytics: these services usually perform complex and time-consuming computations on big data. Examples of workloads for testing offline services are machine learning algorithms such as *k-means clustering* and *naive Bayes classification*. (3) Real-time analytics: application users use these services in an interactive manner; that is, a variety of interactions happen between users and application services. Examples of workloads for these services are relational queries such as *selecting*, *joining*, and *aggregation* of database tables.

# 5   Open Challenges

Considering the emergence of new big data applications and the rapid evolution of big data systems, we believe an incremental and iterative approach is necessary to conduct the investigations on big benchmarks. We now propose some challenges to be addressed to develop successful and efficient big data benchmarks.

## 5.1   Data-Centric Big Data Benchmarks

The fundamental problem of big data benchmarks is about how to provide better measurement of systems for processing data with 4 V properties, which brings the requirement for data-centric benchmarks.

**Table 2.** Comparison of benchmarking techniques.

| Benchmark efforts | Workloads | | Software stacks |
|---|---|---|---|
| | Type | Examples | |
| Hibench [13] | Offline analytics | Sort, WordCount, TeraSort, PageRank, K-means, Bayes classification | Hadoop and Hive |
| | Real-time analytics | Nutch Indexing | |
| GridMix [2] | Online services | Sort, sampling a large dataset | Hadoop |
| PigMix [4] | Online services | 12 data queries | Hadoop |
| YCSB [10] | Online services | OLTP (read, write, scan, update) | NoSQL systems |
| SIGMOD benchmark [16] | Online services | Data loading, select, aggregate, join, count URL links | DBMS and Hadoop |
| TPC-DS [12] | Online services | Data loading, queries and maintenance | DBMS |
| BigBench [12] | Online services | Database operations (select, create and drop tables) | DBMS and Hadoop |
| | Offline analytics | K-means, classification | |
| LinkBench [8] | Online services | Simple operations such as select, insert, update, and delete; and association range queries and count queries | DBMS |
| CloudSuite [11] | Online services | YCSBs workloads | NoSQL systems, Hadoop, GraphLab |
| | Offline analytics | Text classification, WordCount | |
| BigDataBench [19] | Online services | Database operations (read, write, scan) | NoSQL systems, DBMS, real-time and offline analytics systems |
| | Offline analytics | Micro Benchmarks (sort, grep, WordCount, CFS); search engine (index, PageRank); social network (K-means, connected components (CC)); e-commerce (collaborative filtering (CF), Naive Bayes) | |
| | Real-time analytics | Relational database query (select, aggregate, join) | |

**Fully controllable data velocity.** The full control of data velocity has two meanings. First, existing big data benchmarks only consider different data generation rates. Hence different data updating frequencies and processing speeds should be reflected in future big data generators. Secondly, current data velocity is implemented using parallel strategies; that is, data velocity can be controlled by deploying different numbers of parallel data generators. In contrast, we note that data velocity can be controlled in another way: adjusting the efficiency of the data generation algorithms themselves to control data velocity. For example, a graph data generator can be adjusted to consume more memory resources, thus increasing its data generation speed.

**Metrics to evaluate data veracity.** As discussed in Sect. 3.2, applying data models to capture and preserve important characteristics of real data is an efficient way to keep data veracity in synthetic data generation. However, how to measure the conformity of the generated synthetic data to the raw data is still an open question; that is, metrics need to be developed to evaluate data veracity. Two types of evaluation metrics can be developed: (1) metrics to compare the raw data and the constructed data models; (2) metrics to compare the raw data and the synthetic data.

This problem is compounded when considering different data sources and data types. For example, to compare real text data set and synthetic data, we first need to derive the topic and word distributions from these data sets. Next, statistical metrics such as KullbackCLeibler divergence can be applied to compare the similarity between two distributions. Furthermore, when considering table, graph or even stream data, some other metrics should be developed.

## 5.2   Domain-Centric Big Data Benchmarks

The fast development of big data systems has lead to a number of successful application domains such as scientific analytics, social network, and streaming process. Each of these application domain is the focus of one or multiple big data platform efforts. Domain-centric benchmarks, therefore, are needed to promote the progress of these big data platforms.

**Enriching workloads of big data benchmarks.** At present, there are three major problems that restrict the wide application of current big data benchmarks. First, there are still many important big data systems such as multimedia systems and applications such as large-scale deep learning algorithms not being considered. Second, in an application domain, a representative workload should reflect both typical data processing operations and the arrival patterns of these operations (i.e. the arriving rate and sequence of operations). We believe profiling history logs of real applications is a good way to obtain the representative arrival patterns. Finally, the truly hybrid workload, i.e. the workload consists of the mix of various data processing operations and their arriving rates and sequences, has not been adequately supported. That is, to the best of our knowledge, none of exiting big data benchmarks is ready to declare itself to be a truly representative and comprehensive big data benchmark until its workloads are significantly enriched to solve the above problems.

**Supporting heterogeneous hardware platforms.** With the fast development of technology, the emerged hardware platforms and systems significantly change the way about how to process data and show a promising prospect to improve processing efficiency. For example, the heterogeneous platforms of Xeon+General-purpose computing on graphics processing units (GPGPU) and Xeon+Many Integrated Core (MIC) can significantly improve the processing speed of HPC applications. However, to date, both platforms are only limited to the HPC area; that is, the diversity of big data applications are not fully considered in these platforms. For such an issue, big data benchmarks should be developed to evaluate and compare different workloads in state-of-the-practice heterogeneous platforms. The evaluation result is expected to show: (1) whether any platform can consistently win in terms of both performance and energy efficiency for all big data applications, and (2) for each class of big data applications, we hope to find some specific platform that can realize better performance and energy efficiency for them. To support the evaluation of an application, current big data benchmarks should be extended to provide a uniform interface to enable this application running in different platforms. In order to perform apples-to-apples comparisons, this application should be running in the same software stack.

**A reusable environment to automate test generation.** Section 3.3 proposes a framework to abstract operations and workload patterns from typical data processing behaviors, thus enabling automatic generation of tests with comprehensive workloads for big data systems. We note that these operations and patterns are easy to derive in some application domains. For example, in the application domain of basic database operations, there are some obvious operations such as real, write, select, and delete, and the patterns used to combine these operations are simple. However, in some application domains such as social network, there are a large number of data processing operations such as k-means clustering and collaborative filtering, and the relationships between these operations are complex. All these facts mean abstracting a comprehensive set of operations and behaviors is difficult.

Moreover, in practice, generating benchmarking tests from operations and patterns may be beyond the capabilities of the average system owner. Hence going mainstream with this framework requires the development of an environment that provides abstracted operations and workload patterns for different application domains, as well as offers a repository of reusable prescriptions to simplify the generation of prescribed tests running on state-of-the-art software stacks.

**Increasing use cases of big data benchmarks.** The number of successful use cases is an important measure of the practicality of a big data benchmark. Table 3 lists the use cases of current big data benchmarks. We can observe that some benchmarks such as Hibench, GridMix, PigMix are designed to test some specific

Table 3. Use cases of current big data benchmarks.

| Benchmark efforts | Use cases |
| --- | --- |
| Hibench [13], GridMix [2], PigMix [4] | Hadoop |
| YCSB [10] | NoSQL and SQL databases, cloud storage system |
| TPC-DS [12] | DBMS |
| SIGMOD benchmark [16], BigBench [12] | DBMS and Hadoop |
| LinkBench [8] | DBMS (MySQL) |
| CloudSuite [11] | Cloud service architecture |
| BigDataBench [19] | Hardware systems, cloud service architecture, DBMS, distributed systems |

type of systems, hence these benchmarks have a single use case such as Hadoop. In addition, although some benchmarks such as BigBench and LinkBench are designed for different systems, they are currently only evaluated in one system. By contrast, YCSB and BigDataBench have been successfully applied to evaluate multiple systems and architectures. Hence, it is necessary to apply benchmarks to evaluate a larger number and type of big data systems, and in turn, learning from the experiences in testing these systems to develop appropriate benchmarks for better evaluation.

## 6   Conclusion

With the rapid development of information technology, big data systems have emerged to manage and process data with high volume, velocity, and variety. These new systems have given rise to various new requirements about how to develop a new generation of big data benchmarks. In this paper, we summarize the lessons we have learned and propose key challenges in developing big data benchmarks from two aspects: (1) how to develop data generators capable of preserving the 4 V properties of big data in data generation; (2) how to automatically generate benchmarking tests to cover a diversity of typical application scenarios while supporting different system implementations and software stacks. We then introduce the methodology on big data benchmark aiming to address the proposed challenges. Next, we discuss existing benchmarking techniques and propose some future research directions. The work presented in this paper represents our effort towards building a truly representative and comprehensive big data benchmark suite and we encourage more investigations and developments in big data benchmarking tools.

## References

1. Big data benchmark by amplab of uc berkeley (2013). https://amplab.cs.berkeley.edu/benchmark/
2. Gridmix (2013). https://hadoop.apache.org/docs/r1.2.1/gridmix.html
3. Ibm big data platform (2013). http://www-01.ibm.com/software/data/bigdata/
4. Pigmix (2013). https://cwiki.apache.org/confluence/display/PIG/PigMix
5. Sort benchmark (2013). http://sortbenchmark.org/
6. Standard performance evaluation corporation (spec) (2013). http://www.spec.org/gwpg/wpc.static/wpcv1info.html
7. Tpc transaction processing performance council (2013). http://www.tpc.org/
8. Armstrong, T.G., Ponnekanti, V., Borthakur, D., Callaghan, M.: Linkbench: a database benchmark based on the facebook social graph. In: Proceedings of the 2013 International Conference on Management of Data, pp. 1185–1196. ACM (2013)
9. Blei, D.M., Ng, A.Y., Jordan, M.I.: Latent dirichlet allocation. J. Mach. Learn. Res. **3**, 993–1022 (2003)
10. Cooper, B.F., Silberstein, A., Tam, E., Ramakrishnan, R., Sears, R.: Benchmarking cloud serving systems with YCSB. In: Proceedings of the 1st ACM Symposium on Cloud Computing, pp. 143–154. ACM (2010)

11. Ferdman, M., Adileh, A., Kocberber, O., Volos, S., Alisafaee, M., Jevdjic, D., Kaynak, C., Popescu, A.D., Ailamaki, A., Falsafi, B.: Clearing the clouds: A study of emerging workloads on modern hardware. Technical report (2011)

12. Ghazal, A., Rabl, T., Hu, M., Raab, F., Poess, M., Crolotte, A., Jacobsen, H.A.: Bigbench: towards an industry standard benchmark for big data analytics. In: Proceedings of the 2013 International Conference on Management of Data, pp. 1197–1208. ACM (2013)

13. Huang, S., Huang, J., Dai, J., Xie, T., Huang, B.: The hibench benchmark suite: Characterization of the mapreduce-based data analysis. In: 2010 IEEE 26th International Conference on Data Engineering Workshops (ICDEW), pp. 41–51. IEEE (2010)

14. Jia, Z., Wang, L., Zhan, J., Zhang, L., Luo, C.: Characterizing data analysis workloads in data centers. In: 2013 IEEE International Symposium on Workload Characterization (IISWC), pp 66–76. IEEE (2013)

15. Ming, Z., Luo, C., Gao, W., Han, R., Yang, Q., Wang, L., Zhan, J.: Bdgs: A scalable big data generator suite in big data benchmarking. In: Rabl, T., et al. (eds.) Advancing Big Data Benchmarks. LNCS, pp. 138–154. Springer, Heidelberg (2014)

16. Pavlo, A., Paulson, E., Rasin, A., Abadi, D.J., DeWitt, D.J., Madden, S., Stonebraker, M.: A comparison of approaches to large-scale data analysis. In: Proceedings of the 2009 ACM SIGMOD International Conference on Management of Data, pp. 165–178. ACM (2009)

17. Rabl, T., Frank, M., Sergieh, H.M., Kosch, H.: A data generator for cloud-scale benchmarking. In: Nambiar, R., Poess, M. (eds.) TPCTC 2010. LNCS, vol. 6417, pp. 41–56. Springer, Heidelberg (2011)

18. Tay, Y.: Data generation for application-specific benchmarking. VLDB, Challenges and Visions (2011)

19. Wang, L., Zhan, J., Luo, C., Zhu, Y., Yang, Q., He, Y., Gao, W., Jia, Z., Shi, Y., Zhang, S., et al.: Bigdatabench: A big data benchmark suite from internet services. In: Proceedings of the 20th IEEE International Symposium On High Performance Computer Architecture (HPCA-2014), IEEE (2014)

20. Zhu, Y., Zhan, J., Weng, C., Nambiar, R., Zhang, J., Chen, X., Wang, L.: BigOP: generating comprehensive big data workloads as a benchmarking framework. In: Bhowmick, S.S., Dyreson, C.E., Jensen, C.S., Lee, M.L., Muliantara, A., Thalheim, B. (eds.) DASFAA 2014, Part II. LNCS, vol. 8422, pp. 483–492. Springer, Heidelberg (2014)

# A Micro-benchmark Suite for Evaluating Hadoop MapReduce on High-Performance Networks

Dipti Shankar[✉], Xiaoyi Lu, Md. Wasi-ur-Rahman, Nusrat Islam,
and Dhabaleswar K. (DK) Panda

Department of Computer Science and Engineering,
The Ohio State University, Columbus, USA
{shankard,luxi,rahmanmd,islamn,panda}@cse.ohio-state.edu

**Abstract.** Hadoop MapReduce is increasingly being used by many data-centers (e.g. Facebook, Yahoo!) because of its simplicity, productivity, scalability, and fault tolerance. For MapReduce applications, achieving low job execution time is critical. Since a majority of the existing clusters today are equipped with modern, high-speed interconnects such as InfiniBand and 10 GigE, that offer high bandwidth and low communication latency, it is essential to study the impact of network configuration on the communication patterns of the MapReduce job. However, a standardized benchmark suite that focuses on helping users evaluate the performance of the stand-alone Hadoop MapReduce component is not available in the current Apache Hadoop community. In this paper, we propose a micro-benchmark suite that can be used to evaluate the performance of stand-alone Hadoop MapReduce, with different intermediate data distribution patterns, varied key/value sizes, and data types. We also show how this micro-benchmark suite can be used to evaluate the performance of Hadoop MapReduce over different networks/protocols and parameter configurations on modern clusters. The micro-benchmark suite is designed to be compatible with both Hadoop 1.x and Hadoop 2.x.

**Keywords:** Big data · Hadoop MapReduce · Micro-benchmarks · High-performance networks

## 1  Introduction

MapReduce, proposed by Google [8], has been seen as a viable model for processing petabytes of data. The Apache Hadoop project [23], an open-source implementation of the MapReduce computing model, has gained widespread acceptance and is widely used in many organizations around the world. MapReduce is extensively

---

This research is supported in part by National Science Foundation grants #OCI-1148371, #CCF-1213084 and #OCI-1347189. It used the Extreme Science and Engineering Discovery Environment (XSEDE), which is supported by National Science Foundation grant number #OCI-1053575.

© Springer International Publishing Switzerland 2014
J. Zhan et al. (Eds.): BPOE 2014, LNCS 8807, pp. 19–33, 2014.
DOI: 10.1007/978-3-319-13021-7_2

adopted by various applications to perform massive data analysis and is hence required to deliver high performance. While Hadoop does attempt to minimize the movement of data in the network, there are times when MapReduce does generate considerable network traffic, especially during the intermediate data shuffling phase [17], which is communication intensive.

Several modern, high-speed interconnects such as InfiniBand and 10 GigE, are used widely in clusters today. The data shuffling phase of the MapReduce job can immensely benefit from the high bandwidth and low latency communication offered by these high-performance interconnects. In order to evaluate this improvement potential, we require benchmarks that can give us insights into the factors that affect MapReduce as an independent component. The performance of Hadoop MapReduce is influenced by many factors such as network configuration of the cluster, controllable parameters in software (e.g. number of maps/reduces, data distribution), data types, and so on. To get optimal performance, it is necessary to tune and optimize these factors, based on cluster and workload characteristics. Adopting a standardized performance benchmark suite to evaluate these performance metrics in different configurations would be good for Hadoop users. For Hadoop developers, a benchmark suite with these capabilities could help evaluate the performance of new MapReduce designs.

At present, we lack a standardized benchmark suite that focuses on helping users evaluate the performance of the Hadoop MapReduce as a stand-alone component. Current, commonly used benchmarks in Hadoop, such as Sort and Tera-Sort, usually require the involvement of HDFS. The performance of the HDFS component has significant impact on the overall performance of the MapReduce job, and this interferes in the evaluation of the performance benefits of new designs for MapReduce. Furthermore, these benchmarks do not provision us to study the impact of changing data distribution patterns, varying data types, etc., on the performance of the MapReduce job. Such capabilities are very useful for optimizing the parameters and the internal designs of Hadoop MapReduce. With this as background, the basic motivation of this paper is: *Can we design a simple micro-benchmark suite to let users and developers in the Big Data community evaluate, understand, and optimize the performance of Hadoop MapReduce in a stand-alone manner over different networks/protocols?*

In this paper, we propose a comprehensive micro-benchmark suite to evaluate the performance of stand-alone Hadoop MapReduce. We provide options for varying different benchmark-level parameters such as intermediate data distribution pattern, key/value size, data type, etc. Our micro-benchmark suite can also dynamically set the Hadoop MapReduce configuration parameters, like number of map and reduce tasks, etc. We display the configuration parameters and resource utilization statistics for each test, along with the final job execution time, as the micro-benchmark output.

This paper makes the following key contributions:

1. Designing a micro-benchmark suite to evaluate the performance of stand-alone Hadoop MapReduce, when run over different types of high-performance networks.

2. A set of micro-benchmarks to measure the job execution time of Hadoop MapReduce with different intermediate data distribution patterns.
3. Illustration of the performance results of Hadoop MapReduce using these micro-benchmarks over different networks/protocols (1 GigE/10 GigE/IPoIB QDR (32 Gbps)/IPoIB FDR (56 Gbps)).
4. A case study on enhancing Hadoop MapReduce design by using RDMA over native InfiniBand, undertaken with the help of the proposed micro-benchmark suite.

The rest of the paper is organized as follows. In Sect. 2, we discuss related work in the field. We present design considerations for our micro-benchmark suite in Sect. 3 and the micro-benchmarks in Sect. 4. In Sect. 5, we present the results of performance tests, obtained with our micro-benchmark suite. Section 6 shows a case study with RDMA-enhanced MapReduce. Finally, we conclude the paper in Sect. 7.

## 2  Related Work

Over the years, many benchmarks have been introduced in the areas of Cloud Computing and Big Data. MRBench [13] provides micro-benchmarks in the form of MapReduce jobs of TPC-H [4]. MRBS [21] is a benchmark suite that evaluates the dependability of MapReduce systems. It provides five benchmarks for several application domains and a wide range of execution scenarios. Similarly, HiBench [9] has extended the DFSIO program to compute the aggregated throughput by disabling the speculative execution of the MapReduce framework. It also evaluates Hadoop in terms of system resource utilization (e.g. CPU, memory). MalStone [6] is a benchmark suite designed to measure the performance of cloud computing middleware when building data mining models. Yahoo! Cloud Serving Benchmark (YCSB) [7] is a set of benchmarks for performance evaluations of key/value-pair and cloud data-serving systems. YCSB++ [18] further extends YCSB to improve performance understanding and debugging. BigData-Bench [25], a benchmark suite for Big Data Computing, covers typical Internet service workloads and provides representative data sets and data generation tools. It also provides different implementations for various Big Data processing systems [1,14].

In addition to the above benchmarks that address the Hadoop framework as a whole, micro-benchmark suites have been designed to study some of its individual components. The micro-benchmark suite designed in [10] helps with detailed profiling and performance characterization of various HDFS operations. Likewise, the micro-benchmark suite designed in [16] provides detailed profiling and performance characterization of Hadoop RPC over different high-performance networks. Along these lines, our proposed micro-benchmark suite introduces a performance evaluation tool for stand-alone Hadoop MapReduce, that does not need HDFS or any other distributed file system.

## 3  Design Considerations

The performance of a MapReduce job is usually measured by its execution time. It can be significantly influenced by numerous factors such as the underlying network or the communication protocol, number of map tasks and reduce tasks, intermediate shuffle data pattern, and the shuffle data size, as illustrated in Fig. 1(a).

Essentially, the efficiency of the network intensive data shuffling phase is determined by how fast the map outputs are shuffled. Based on these aspects, we consider the following dimensions to design the Hadoop MapReduce micro-benchmark suite,

**Intermediate data distribution:** Map tasks transform input key/value pairs to a set of intermediate key/value pairs. These intermediate key/value pairs are shuffled from the mappers, where they are created, to the reducers where they are consumed. Depending upon the MapReduce job, the distribution of intermediate map output records can be even or skewed. A uniformly balanced load can significantly shorten the total run time by enabling all reducers to finish at about the same time. In jobs with a skewed load, some reducers complete the job quickly, while others take much longer, as the latter have to process a more sizeable portion of the work. Since it is vital for performance to understand whether these distributions can be significantly impacted by the underlying network protocol, we consider this an important aspect of our micro-benchmark suite.

**Size and number of key/value pairs:** For a given data size, the size of the key/value pair determines the number of times the Mapper and Partitioner functions are called; and, in turn, the number of intermediate records being shuffled. Our micro-benchmark suite provides support for three related parameters: key size, value size and number of key/value pairs. Through these parameters, we can specify the total amount of data to be processed by each map, amounting to the total shuffle data size. These parameters can help us understand how the nature of intermediate data, such as the key/value pair sizes, can impact the performance of the MapReduce job, on different networks.

**Number of map and reduce tasks:** The number of map and reduce tasks is probably the most basic Hadoop MapReduce parameters. Tuning the number of map and reduce tasks for a job is essential for optimal performance and hence we provide support to vary the number of map and reduce tasks in our micro-benchmark suite.

**Data types:** Hadoop can process many different types of data formats, from flat text files to databases. Binary data types usually take up less space than textual data. Since disk I/O and network transfer will become bottlenecks in large jobs, reducing the sheer number of bytes taken up by the intermediate data can provide a substantial performance gain. Thus, data types can have a considerable impact on the performance of the MapReduce job. Our micro-benchmark suite is designed to support different data types.

**Network configuration:** The intermediate data shuffling phase, which is the heart of MapReduce, results in a global, all-to-all communication. This accounts for a rather significant amount of network traffic within the cluster, although it varies from job to job. This is an important consideration, especially when expanding the cluster. Hence, it is essential to compare and contrast the impact that different network interconnects/protocols have on the performance of the MapReduce job. Our micro-benchmark suite is capable of running over any network and cluster configuration.

**Resource utilization:** The multi-phase parallel model of MapReduce and its scheduling policies have a significant impact on various systems resources such as the CPU, the memory, and the network, especially with an increasing number of maps and reduce tasks being scheduled. As it is essential to understand the correlation between network characteristics and resource utilization, our micro-benchmark suite provides the capability to measure the resource utilization, during the course of the MapReduce job.

# 4    Micro-benchmarks for Hadoop MapReduce

In this section, we present the overall design of the micro-benchmark suite, and describe the micro-benchmarks implemented based the various design factors outlined in Sect. 3.

## 4.1    Overview of Overall Micro-benchmark Suite Design

In this study, we develop a micro-benchmark suite for stand-alone Hadoop MapReduce, in order to provide an understanding of the impact of different factors described in Sect. 3 on the performance of the MapReduce job when run over different networks. As illustrated in Fig. 1(b), these micro-benchmarks have the following key features:

**Stand-alone feature:** Each micro-benchmark is basically a MapReduce job that is launched without HDFS or any other distributed file system. In order to achieve this,

(1) For the Map phase, we define a custom input format, namely, `NullInputFormat`, for the mapper instances. This empty input format creates dummy input splits based on the number of map tasks specified, with a single record in each. Each map task generates a user-specified number of key/value pairs in memory, and passes it on as map output.

(2) For the Reduce phase, we make use of `NullOutputFormat` [3], defined in the MapReduce API, as the output format. Each reduce task aggregates intermediate data from the map phase, iterates over them and discards it to `/dev/null`. This is ideal for our micro-benchmarks, since we evaluate MapReduce as a stand-alone component.

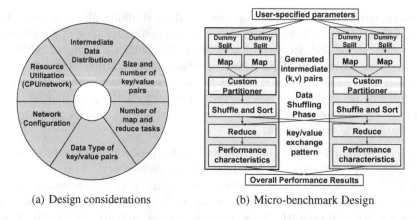

(a) Design considerations          (b) Micro-benchmark Design

**Fig. 1.** Design overview of the MapReduce micro-benchmarks

**Custom Partitioners:** The partitioning phase that takes place after the map phase and before the reduce phase, partitions the data gets across the reducers according to the partitioning function. To simulate different intermediate data distribution scenarios, we employ different custom partitioners. We support three different intermediate data distribution patterns, namely, average distribution, random distribution and skewed distribution. These three distributions cover most of the intermediate data distribution patterns found in MapReduce applications.

**Configurable parameters:** In our micro-benchmark suite, we provide parameters to vary the number of mappers and reducers. The user can specify the size in bytes and the number of the key/value pairs to be generated based on the intermediate shuffle data size of the MapReduce job. We also provide a parameter to indicate data type, such as `BytesWritable` or `Text`. Based on these parameters, the micro-benchmark suite generates the data to be processed by each map.

### 4.2 Micro-benchmarks

Using the framework described in Sect. 4.1 as basis, we define three micro-benchmarks. For each of these, we can vary the size and number of key/value pairs, to generate different sizes of shuffle data. These micro-benchmarks use the `Map` function to create specified number of key/value pairs. To avoid any additional overhead, we restrict the number of unique pairs generated to the number of reducers specified.

**Average-distribution (MR-AVG):** In this micro-benchmark, we distribute the intermediate key/value pairs uniformly amongst all of the reduce tasks. The custom partitioner defined for MR-AVG distributes the key/value pairs amongst the reducers in a round-robin fashion, making sure each reducer gets the same number of intermediate key/value pairs. This helps us obtain a fair comparison of the performance of a MapReduce job on different networks, when the intermediate data is evenly distributed.

**Random-distribution (MR-RAND):** In this micro-benchmark, we randomly distribute the intermediate key/value pairs among the reduce tasks. The custom partitioner defined for MR-RAND randomly picks a reducer and assigns the key/value pair to it. With the help of Java's Random class, the reducer is pseudo-randomly chosen, with its range specified as the number of reducers. With this limited range, the micro-benchmark more or less generates the same pattern of reducers, making sure each run gets similar intermediate key/value pairs to reducers mapping. This mapping is relatively close to an even distribution and thus helps us picture a fairly accurate comparison of the performance of a MapReduce job on different networks, when the intermediate data is randomly distributed among the reducers.

**Skew-distribution (MR-SKEW):** In this micro-benchmark, we distribute the intermediate key/value pairs unevenly among the reducers. Based on the number of reducers specified, the custom partitioner defined for MR-SKEW distributes 50 % of the intermediate key/value pairs to the first reducer, 25 % of the remainder to the second reducer, 12.5 % of the remaining to the third, and then randomly distributes the rest. Since this skewed distribution pattern is fixed for all runs, irrespective of the key/value pairs generated, we can guarantee a fair comparison on homogenous systems. This micro-benchmark helps us gain essential insights into the performance of MapReduce jobs with skewed loads, running over different network types. By determining the overhead of running a skewed load, we can determine if it is worthwhile to find alternative techniques that can mitigate load imbalances in Hadoop applications.

## 5 Performance Evaluation

### 5.1 Experimental Setup

**(1) Intel Westmere Cluster (Cluster A):** This cluster has nine nodes. Each node has Xeon Dual quad-core processor operating at 2.67 GHz. Each node is equipped with 24 GB and two 1TB HDDs. Nodes in this cluster also have NetEffect NE020 10Gb Accelerated Ethernet Adapter that are connected using a 24 port Fulcrum Focalpoint switch. The nodes are also interconnected with a Mellanox switch. Each node runs Red Hat Enterprise Linux Server release 6.1.

**(2) TACC Stampede [22] (Cluster B):** We use the Stampede supercomputing system at TACC [22] for our experiments. According to TOP500 [24] list in June 2014, this cluster is listed as the 7th fastest supercomputer worldwide. Each node in this cluster is dual socket containing Intel Sandy Bridge (E5-2680) dual octa-core processors, running at 2.70GHz. It has 32 GB of memory, a SE10P (B0-KNC) co-processor and a Mellanox IB FDR MT4099 HCA. The host processors are running CentOS release 6.3 (Final). Each node is equipped with a single 80 GB HDD.

For Cluster A, we show performance comparisons over 1 GigE, 10 GigE, and IPoIB (32 Gbps). We evaluate our micro-benchmarks with Apache Hadoop 1.2.1 and JDK version 1.7. We also present some results with Apache Hadoop NextGen MapReduce (YARN) [5] version 2.4.1. On Cluster B, we compare IPoIB FDR (56 Gbps) and RDMA FDR (56 Gbps), with Apache Hadoop 1.2.1 running over IPoIB and RDMA for Apache Hadoop (v0.9.9) [2]. These results are presented in Sect. 6.

## 5.2  Evaluation along Different Dimensions

**Evaluating impact of intermediate data distribution patterns:** In this section, we present performance results obtained using micro-benchmarks presented in Sect. 4.2, to evaluate the impact of intermediate data distribution patterns on the MapReduce job execution time. We perform these tests on Cluster A, using BytesWritable data type and a fixed key/value pair size of 1 KB, with 16 map tasks and 8 reduce tasks on 4 slave nodes. We compare the performance with different shuffle data sizes, by varying the number of intermediate key/value pairs generated. Figure 2 shows comparison for job execution time with different intermediate data distribution patterns. From Fig. 2(a), it is clear that the job execution time for MR-AVG micro-benchmark decreases around 17 %, if the underlying interconnect is changed to 10 GigE from 1 GigE, and up to 24 %, when changed to IPoIB (32 Gbps). Similarly, from Fig. 2(b), job execution time for MR-RAND micro-benchmark decreases around 16 %, if the underlying interconnect is 10 GigE instead of 1 GigE, and up to 22 %, if it is IPoIB (32 Gbps). We observe that IPoIB (32 Gbps) improves the performance by about 8–10 %, as compared to 10 GigE, for both MR-AVG and MR-RAND micro-benchmarks. From Fig. 2(c), we can observe that the performance improves by about 11 % for MR-SKEW micro-benchmark, if we switch from 1 GigE from 10 GigE. Also, IPoIB (32 Gbps) performs better than 10 GigE, by about 12 %, as intermediate shuffle data sizes are scaled up. It can be observed that IPoIB (32 Gbps) provides better improvement with increased shuffle data sizes and more skewed workloads. Also, the skewed data distribution seems to double the job execution time for a given data size, as compared to the average distribution, irrespective of the underlying network interconnect.

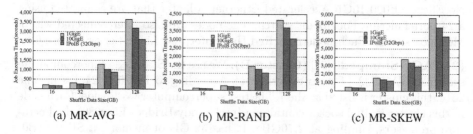

(a) MR-AVG          (b) MR-RAND          (c) MR-SKEW

**Fig. 2.** Job Execution Time for different data distribution patterns on Cluster A

**Evaluating with Apache Hadoop NextGen MapReduce (YARN):** In this section, we evaluate the probable influence that the intermediate data distribution pattern has on the MapReduce job execution time for the Hadoop YARN Architecture [5], using Apache Hadoop 2.4.1. We perform these tests on Cluster A, with a key/value pair size of 1 KB, using 32map tasks and 16 reduce tasks on 8 slave nodes. We compare the performance with different shuffle data sizes, by varying the number of intermediate key/value pairs generated. From Fig. 3(a), it is clear that the job execution time for MR-AVG decreases around 11 %, if the underlying interconnect is changed to 10 GigE from 1 GigE, and by about 18 %, when changed to IPoIB (32 Gbps). Similarly, from Fig. 3(b), job execution time for MR-RAND micro-benchmark decreases around 10 %, when we switch from 1 GigE to 10 GigE, and up to 17 % improvement, when changed to IPoIB (32 Gbps). For MR-SKEW micro-benchmark, Fig. 3(c) shows that the performance of the MapReduce job improves by about 10–12 % with the use of high-speed interconnects. It can be observed that, IPoIB (32 Gbps) improves performance by about 7–10 % for all three micro-benchmarks, as compared to 10 GigE, with increased shuffle data sizes. Also, for a given data size, the skewed data distribution increases the job execution time by more than 3X, when compared to the average data distribution, irrespective of the underlying network interconnect. From Figs. 2 and 3, we can infer that, increasing cluster size and concurrency significantly benefits average and random data distribution patterns. From Figs. 2(c) and 3(c), it also evident that, even though the Map phase may benefit from the increased concurrency and cluster size, the Reduce phase of the MapReduce job with a skewed intermediate data distribution still depends on the slowest reduce task, and hence, the improvement is not as much.

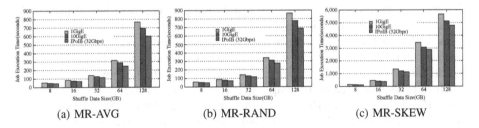

(a) MR-AVG              (b) MR-RAND              (c) MR-SKEW

**Fig. 3.** Job Execution Time with different patterns for YARN architecture on Cluster A

**Evaluating impact of varying key/value pair sizes:** In this section, we present results using MR-AVG micro-benchmark, to portray the impact of varying key/value pair size on the performance of the MapReduce job. We run these evaluations on Cluster A, with 16 map tasks and 8 reduce tasks on 4 slave nodes, for `BytesWritable` data type. Figure 4 shows job execution time comparisons with MR-AVG micro-benchmark for different key/value pair sizes. From Fig. 4(a), we can see that the job execution time for a key/value pair size of 100 bytes, decreases around 18 %, if the underlying interconnect is changed from

1 GigE to 10 GigE, and by about 22 %, when changed to IPoIB (32 Gbps). Figures 4(b) and (c) show similar performance improvement for key/value pair sizes of 1 KB and 100 KB, when underlying interconnect is changed to from 1 GigE to IPoIB (32 Gbps) and 10 GigE. We also observe that IPoIB (32 Gbps) performs slightly better than 10 GigE, by about 7–10 %, for all three key/value pair sizes. It can be seen that increasing the key/value pair sizes brings about lower job execution times for a given shuffle data size. For instance, the job execution time for 16 GB shuffle data size reduces from 128 to 107 s for IPoIB (32 Gbps) when key/value sizes are increased from 100 bytes to 10 KB. We can therefore infer that the size and number of key/value pairs can influence the performance of the MapReduce job running on different networks.

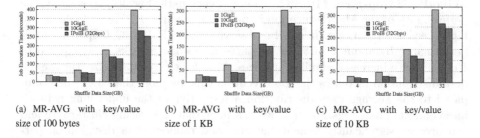

(a) MR-AVG with key/value size of 100 bytes

(b) MR-AVG with key/value size of 1 KB

(c) MR-AVG with key/value size of 10 KB

**Fig. 4.** Job Execution Time with MR-AVG for different key/value pair sizes on Cluster A

**Evaluating impact of varying the number of map and reduce tasks:** In this section, we present performance results with varying number of map and reduce tasks, using MR-AVG micro-benchmark, over 10 GigE and IPoIB (32 Gbps). We run these performance evaluations on Cluster A, with a key/value pair size of 1 KB. We vary the number of key/value pairs to generate different shuffle data sizes. In Fig. 5, we present performance evaluations with 8 map and 4 reduce tasks (8M-4R), and 4 map and 2 reduce tasks (4M-2R). For both these cases, Fig. 5 clearly shows that IPoIB (32 Gbps) outperforms 10 GigE, by about 13 %. It is evident that IPoIB (32 Gbps) gives better performance improvement with increased concurrency, as compared to 10 GigE. For instance, increasing the number of map and reduce tasks improved the performance of the MapReduce job by about 32 % for IPoIB (32 Gbps), while it improved by only 24 % for 10 GigE, for a shuffle data size of 32 GB. It can therefore be inferred that varying the number of map and reduce tasks can impact the load on the network.

**Evaluating impact of data types:** In this section, we present results of experiments done with the MR-RANDOM micro-benchmark, to understand the impact of data types on the performance of the MapReduce job over different networks interconnects. We run these experiments on Cluster A, using 16 map tasks and 8 reduce tasks on 4 slave nodes, with fixed key/value pair size of 1 KB,

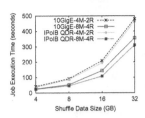

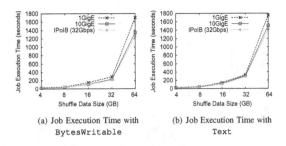

(a) Job Execution Time with
BytesWritable

(b) Job Execution Time with
Text

**Fig. 5.** Job Execution Time
with varying number of maps
and reduces on Cluster A

**Fig. 6.** Job Execution Time with `BytesWri-`
`table` and `Text` on Cluster A

to study `BytesWritable` and `Text` data types. Figure 6(a) shows the trend for
`BytesWritable` and Fig. 6(b) shows the trend for `Text`, as we scale up to 64 GB.
We observe that the job execution time decreases around 17–23 %, if the underlying interconnect is 10 GigE, instead of 1 GigE, and up to 28 % improvement,
if it is IPoIB (32 Gbps). Also, IPoIB (32 Gbps) gives noticeable performance
improvement over 10 GigE. It is evident that high-speed interconnects provide
similar improvement potential to both data types. We plan to investigate other
data types, as the next step.

**Resource Utilization:** In this section, we present results of experiments done
with the MR-AVG micro-benchmark, to study the resource utilization patterns of
MapReduce jobs over different network interconnects. We use 16 map tasks and
8 reduce tasks on 4 slave nodes for these experiments, on Cluster A. We present
CPU and network utilization statistics of one of the slave nodes. Figure 7(a)
shows the CPU utilization trends for MR-AVG benchmark, run with intermediate data size of 16 GB, a fixed key/value pair size of 1 KB and `BytesWritable`
data type. It can be observed that CPU utilization trends of 10 GigE and IPoIB
(32 Gbps) are similar to that of 1 GigE. Figure 7(b) shows the network throughput for the same. For network throughput, we consider the total number of
megabytes received per second. IPoIB (32 Gbps) achieves a peak bandwidth of
950 MB/s, 10 GigE peaks at 520 MB/s, and 1 GigE peaks at 101 MB/s. These
trends suggest that IPoIB (32 Gbps) makes better use of the resources, especially
the network bandwidth, as compared to 10 GigE and 1 GigE.

## 6   A Case Study: Enhanced Hadoop MapReduce Design over Native InfiniBand

In previous research [19,20], we have designed and implemented a high-
performance Hadoop MapReduce framework with RDMA over native Infini-
Band, known as MRoIB. This is publicly available as a part of the RDMA for
Apache Hadoop project (v0.9.9) [2,11,12,15,19,20]. We found that the stand-
alone Hadoop MapReduce micro-benchmark suite proposed in this paper is

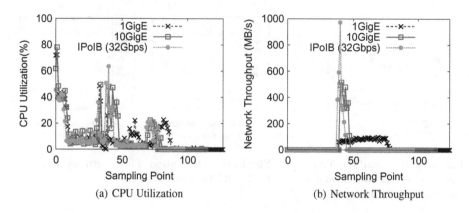

**Fig. 7.** Resource Utilization on one slave node for MR-AVG on Cluster A

extremely helpful in evaluating the performance of alternative MapReduce designs such as MRoIB, and in tuning different internal parameters to obtain optimal performance.

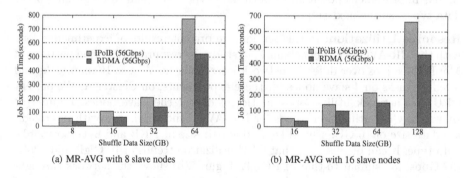

**Fig. 8.** Performance with different patterns for IPoIB Vs. RDMA on Cluster B (56 Gbps FDR)

Figure 8 illustrates the performance improvement possible with native IB as compared to IPoIB (56 Gbps) on Cluster B. We use BytesWritable data type and a fixed key/value pair size of 1 KB, with 32 map tasks and 16 reduce tasks. We vary the number of key/value pairs to generate different shuffle data sizes, and study the MR-AVG micro-benchmark. We omit the other two micro-benchmarks due to space constraints. From Fig. 8(a), we observe that MRoIB improves the performance of the MapReduce job running on 8 slaves nodes, by 28–30 %, as compared to default Hadoop MapReduce over IPoIB (56 Gbps). Similarly, Fig. 8(b) illustrates a comparison between MRoIB and default MapReduce over IPoIB (56 Gbps) with 16 slave nodes on Cluster B. It is clear that RDMA-enhanced MapReduce outperforms IPoIB (56 Gbps) by about 25–28 %, even on a larger cluster. This points us towards the benefits that native IB-based MapReduce has over default Hadoop MapReduce running over IPoIB (56 Gbps).

## 7   Conclusion and Future Work

In order to obtain optimal performance, it is essential to study the impact of network on the performance of Hadoop MapReduce. In this paper, we have designed a micro-benchmark suite to evaluate the performance of stand-alone MapReduce over different network interconnects. This comprehensive and easy-to-use micro-benchmark suite, that is compatible with both Hadoop 1.x and Hadoop 2.x, gives users a means to understand how factors such as intermediate data patterns, size and number of key/value pairs, data type, and number of map and reduce tasks, can influence the execution of a MapReduce job on high-performance networks.

As an illustration, we have presented performance results of Hadoop MapReduce with our micro-benchmarks over different networks/protocols: 1 GigE, 10 GigE, IPoIB QDR (32 Gbps), and IPoIB FDR (56 Gbps). We observe that the performance of the MapReduce job improves around 17 %, if the underlying interconnect is changed to 10 GigE from 1 GigE, and up to 23 %, when changed to IPoIB QDR (32 Gbps). Additionally, IPoIB QDR (32 Gbps) improves performance of the MapReduce job by about 12 % over 10 GigE. It is also noticeable that IPoIB QDR (32 Gbps) performs better with increasing shuffle data sizes. We also present a case study undertaken to understand the benefits that native InfiniBand can provide to Hadoop MapReduce. It is clear that RDMA-enhanced MapReduce design can achieve much better performance than default Hadoop MapReduce over IPoIB FDR (56 Gbps).

In the light of the results presented in this paper, our proposed micro-benchmark suite can help developers enhance their MapReduce designs, especially those intended to optimize data shuffling over the network. For future work, we plan to provide public access to these micro-benchmarks, by making them available as a part of the OSU HiBD Micro-benchmarks [2]. We also intend to add additional features to enhance this micro-benchmark suite, so that users can gain a more concrete understanding of real-world workloads.

## References

1. BigDataBench: A Big Data Benchmark Suite. http://prof.ict.ac.cn/BigDataBench
2. High-Performance Big Data (HiBD). http://hibd.cse.ohio-state.edu
3. NullOutputFormat (Hadoop 1.2.1 API). https://hadoop.apache.org/docs/r1.2.1/api/org/apache/hadoop/mapred/lib/NullOutputFormat.html
4. TPC Benchmark H - Standard Specication. http://www.tpc.org/tpch
5. Apache Hadoop NextGen MapReduce (YARN). http://hadoop.apache.org/docs/current/hadoop-yarn/hadoop-yarn-site/YARN.html
6. Bennett, C., Grossman, R.L., Locke, D., Seidman, J., Vejcik, S.: Malstone: Towards a benchmark for analytics on large data clouds. In: Proceedings of the 16th ACM SIGKDD International Conference on Knowledge Discovery and Data Mining, KDD, Washington, DC, USA (2010)

7. Cooper, B.F., Silberstein, A., Tam, E., Ramakrishnan, R., Sears, R.: Benchmarking cloud serving systems with YCSB. In: Proceedings of the 1st ACM Symposium on Cloud Computing, SoCC, Indianapolis, Indiana, USA (2010)

8. Dean, J., Ghemawat, S.: MapReduce: simplified data processing on large clusters. In: Proceedings of the 6th Conference on Symposium on Opearting Systems Design and Implementation, OSDI, San Francisco, CA (2004)

9. Huang, S., Huang, J., Dai, J., Xie, T., Huang, B.: The HiBench benchmark suite: characterization of the MapReduce-based data analysis. In: Proceedings of the 26th International Conference on Data Engineering Workshops, ICDEW, Long Beach, CA, USA (2010)

10. Islam, N.S., Lu, X., Wasi-ur-Rahman, M., Jose, J., (DK) Panda, D.K.: A micro-benchmark suite for evaluating HDFS operations on modern clusters. In: Rabl, T., Poess, M., Baru, C., Jacobsen, H.-A. (eds.) WBDB 2012. LNCS, vol. 8163, pp. 129–147. Springer, Heidelberg (2014)

11. Islam, N.S., Rahman, M.W., Jose, J., Rajachandrasekar, R., Wang, H., Subramoni, H., Murthy, C., Panda, D.K.: High performance RDMA-based design of HDFS over InfiniBand. In: The International Conference for High Performance Computing, Networking, Storage and Analysis (SC), November 2012

12. Islam, N.S., Lu, X., Rahman, M.W., Panda, D.K.D.: SOR-HDFS: a SEDA-based approach to maximize overlapping in RDMA-enhanced HDFS. In: Proceedings of the 23rd International Symposium on High-Performance Parallel and Distributed Computing, HPDC '14, Vancouver, BC, Canada, pp. 261–264. ACM (2014)

13. Kim, K., Jeon, K., Han, H., Kim, S., Jung, H., Yeom, H.: MRBench: a benchmark for MapReduce framework. In: Proceedings of the IEEE 14th International Conference on Parallel and Distributed Systems, ICPADS, Melbourne, Victoria, Australia (2008)

14. Liang, F., Feng, C., Lu, X., Xu, Z.: Performance benefits of DataMPI: a case study with BigDataBench. In: The 4th Workshop on Big Data Benchmarks, Performance Optimization, and Emerging Hardware, BPOE-4, Salt lake, Utah (2014)

15. Lu, X., Islam, N.S., Rahman, M.W., Jose, J., Subramoni, H., Wang, H., Panda, D.K.: High-performance design of hadoop RPC with RDMA over InfiniBand. In: Proceedings of the IEEE 42th International Conference on Parallel Processing, ICPP, Lyon, France (2013)

16. Lu, X., Islam, N.S., Wasi-Ur-Rahman, M., Panda, D.K.: A Micro-benchmark suite for evaluating hadoop RPC on high-performance networks. In: Proceedings of the 3rd Workshop on Big Data Benchmarking, WBDB (2013)

17. Lu, X., Wang, B., Zha, L., Xu, Z.: Can MPI benefit hadoop and MapReduce applications? In: Proceedings of the IEEE 40th International Conference on Parallel Processing Workshops, ICPPW (2011)

18. Patil, S., Polte, M., Ren, K., Tantisiriroj, W., Xiao, L., López, J., Gibson, G., Fuchs, A., Rinaldi, B.: YCSB++: benchmarking and performance debugging advanced features in scalable table stores. In: Proceedings of the 2nd ACM Symposium on Cloud Computing, SoCC, Cascais, Portugal (2011)

19. Rahman, M.W., Islam, N.S., Lu, X., Jose, J., Subramoni, H., Wang, H., Panda, D.K.: High-Performance RDMA-based Design of Hadoop MapReduce over Infini-Band. In: Proceedings of the IEEE 27th International Symposium on Parallel and Distributed Processing Workshops and PhD Forum. IPDPSW, Washington, DC, USA (2013)

20. Rahman, M.W., Lu, X., Islam, N.S., Panda, D.K.: HOMR: a hybrid approach to exploit maximum overlapping in MapReduce over high performance interconnects. In: Proceedings of the 28th ACM International Conference on Supercomputing, ICS '14, Munich, Germany, pp. 33–42. ACM (2014)
21. Sangroya, A., Serrano, D., Bouchenak, S.: MRBS: towards dependability benchmarking for hadoop MapReduce. In: Caragiannis, I., et al. (eds.) Euro-Par 2012 Workshops 2012. LNCS, vol. 7640, pp. 3–12. Springer, Heidelberg (2013)
22. Stampede at Texas Advanced Computing Center. http://www.tacc.utexas.edu/resources/hpc/stampede
23. The Apache Software Foundation: Apache Hadoop. http://hadoop.apache.org
24. Top500 Supercomputing System. http://www.top500.org
25. Wang, L., Zhan, J., Luo, C., Zhu, Y., Yang, Q., He, Y., Gao, W., Jia, Z., Shi, Y., Zhang, S., Zheng, C., Lu, G., Zhan, K., Li, X., Qiu, B.: BigDataBench: a big data benchmark suite from internet services. In: Proceedings of the 20th IEEE International Symposium on High Performance Computer Architecture, HPCA, Orlando, Florida (2014)

# MemTest: A Novel Benchmark
# for In-memory Database

Qiangqiang Kang, Cheqing Jin$^{(\boxtimes)}$, Zhao Zhang, and Aoying Zhou

Institute for Data Science and Engineering,
Software Engineering Institute, East China Normal University, Shanghai, China
qqkang@ecnu.cn, {cqjin,zhzhang,ayzhou}@sei.ecnu.edu.cn

**Abstract.** With the rapid development of hardware, a stand-alone computer can employ a memory which has large amounts of volumes. Several industries and research institutions have devoted more resources to develop several in-memory databases, which preload the data into memory for further processing. With the boom of in-memory databases, there emerges requirements to evaluate and compare the performance of these systems impartially and objectively. In this paper, we proposed MemTest, a novel benchmark considering the main characteristics of an in-memory database. This benchmark constructs particular metrics, which cover CPU usage, cache miss, compression ratio, minimal memory space and response time of an in-memory database and are also the core of our benchmark. We design a data model based on inter-bank transaction applications, around which a data generator is devised to support the data distributions of uniform and skew. The MemTest workload includes a set of queries and transactions against the metrics and data model. In the end, we illustrate the efficacy of MemTest through implementations on three different in-memory databases.

**Keywords:** Memory · In-memory database · Benchmark · Finance

## 1 Introduction

Nowadays, the price of memory continues to decrease, making it possible to deploy a computer system with huge memory size, and chip densities continue their current trend of doubling every year for the foreseeable future [1]. Several commercial and open source providers have devoted huge resources to develop IMDBs (In-Memory Databases), such as Hana (SAP) [2], Timesten (Oracle) [3], Hekaton (Microsoft) [4], HyperSql [5], SQLite [6], MemSql [7] and Monetdb (OpenSource) [8]. In general, IMDBs preload the whole data into memory so that the I/O operators can be significantly avoided during the query processing. Hence, it is expected that IMDBs will play an important role in some emergent applications, such as weather forecast, finance, artificial intelligence, etc. The appearance of more and more IMDB products brings a need to devise a benchmark to test and evaluate them fairly and objectively.

© Springer International Publishing Switzerland 2014
J. Zhan et al. (Eds.): BPOE 2014, LNCS 8807, pp. 34–46, 2014.
DOI: 10.1007/978-3-319-13021-7_3

The development of benchmark has shown great success in the past 30 years, since the birth of Wisconsin Benchmark. Academia and industry have proposed numerous meaningful database benchmarks, such as the Wisconsin benchmark and TPC-X series for RDBMS, the OO7 [9] and bucky [10] for object-oriented DBMS, XMark [11] and EXRT [12] for XML data, YCSB [13] and BigBench [14] for big data applications, etc. Existing benchmarks are often employed to test the performance for the disk-based databases, regarding throughput and response time as main metrics. Such benchmarks cannot be suitable for IMDBs well since the following characters of an IMDB are not addressed well [2,3,8,15]: (i) Data Compression (In IMDBs, data is often stored in compressed form. Optimizations for this characteristic can reduce the data size in main memory. Hence, it is required to notice this property.), (ii) Minimal Memory Space (Memory is critical in IMDBs. During the query processing, it not only stores the data but also offers enough space to process data. Each IMDB has a request for the minimal memory space.), (iii) CPU (In database fields, CPU is used for computation and data processing. Different from most conventional systems which attempt to minimize disk access, IMDBs have overlooked I/O cost and focus more on the processing cost of CPU. By the optimization of CPU utility, an IMDB can speed up the query processing and gain more performance improvement) and (iv) Cache (In IMDBs, cache is employed to reduce the data transfer between memory and cpu. Actually, during the data processing, performance of CPU depends upon how well the cache can be utilized.).

Our goal in this paper is a novel benchmark for in-memory databases, named MemTest. Based on main properties of IMDBs, some metrics are proposed to evaluate the performance fairly and objectively, including compression ratio, response time, minimal memory space, CPU usage and cache miss. For other parts, the benchmark embraces a data model, synthetic data generator and workload description.

The data model in MemTest is based on an inter-bank transaction scenario. This type of applications needs to capture, store, manage and analyze terabytes of data every day. In order to satisfy daily work, IMDBs are employed as a innovative solution to process the huge data. Many companies works on the application, such as VISA, JCB and China Unionpay. In this study, we employ a star schema including six tables to construct the schema. There are one big table as the fact table (having more than 200 columns) and five small tables as dimension tables (less than 30 columns). The data types cover text, date and numeric, etc. Based on realistic data properties, we provide the implementation of a simple data generator to generate dataset with uniform distribution and skew distribution. A scale factor (SF) is also provided to decide scalable volumes of raw data. Then, we devised 12 queries and 2 transactions in the workload part. Six major business areas are identified: institutions' transaction statistics and analysis for each day, transaction quality analysis, transaction compliance analysis, institutions' abnormal statistics and analysis, generating new transactions and capturing the most abnormal institutions. The workload supports join, aggregation, and update, etc. in the IMDB field.

The major contribution can be summarized by Table 1, showing that our proposed benchmark actually deals with some intrinsic characters of an IMDB system. As well as the existing benchmarks, our novel benchmark can cover important characteristics in DBMS, such as *OLTP, OLAP* and *Multi-user model*. Meanwhile, MemTest still takes special consideration of some inherent characteristics of the IMDB systems, most of which are often overlooked by the existing works. As mentioned above, several characteristics, such as *compression ratio, startup, Minimal Memory Space*, have not been studied before. In addition, we also conduct a series of experiments to evaluate IMDBs.

**Table 1.** The comparison of MemTest and other benchmarks

| Characteristic | SSB | TPC-C | TPC-H | TPC-DS | CH-Benchmark | MemTest |
|---|---|---|---|---|---|---|
| OLTP | ✗ | ✓ | ✗ | ✗ | ✓ | ✓ |
| OLAP | ✓ | ✗ | ✓ | ✓ | ✓ | ✓ |
| Multi-user model | ✗ | ✓ | ✓ | ✓ | ✓ | ✓ |
| Compression Ratio | ✗ | ✗ | ✗ | ✗ | ✗ | ✓ |
| Minimal Memory Space | ✗ | ✗ | ✗ | ✗ | ✗ | ✓ |
| CPU Usage | ✗ | ✗ | ✗ | ✗ | ✗ | ✓ |
| Cache Miss | ✗ | ✗ | ✗ | ✗ | ✗ | ✓ |

The rest of this paper is organized as follows. Section 2 describes the related work. Section 3 emphasizes on the proposed metrics. Section 4 illustrates the data model and its data generation. Section 5 introduces the detailed world. We report some experimental results in Sect. 6, and conclude this work briefly in the last section.

## 2    Related Work

The requirement for well-defined benchmarks that measure the performance of database has grown with time passing by. Since 1980's, academia and industry have proposed numerous meaningful database benchmarks, such as the Wisconsin benchmark and TPC-X series for RDBMS, the OO7 [9] and bucky [10] for object-oriented DBMS, XMark [11] and EXRT [12] for XML data, YCSB [13] and BigBench [14] for big data applications, etc.

With the rapid development of hardware in recent years, we can deploy huge amounts of RAM in one computer system, which makes in-memory database (IMDB) possible. Representative IMDB products include Hana (SAP), MonetDB, Timesten (Oracle), HyperSql, SQLite and MemSql, all based on relational data model. The benchmarks for RDBMS have made big progress in the past 30 years. Wisconsin benchmark [16] is the earliest benchmark. It develops a relatively simple, but fairly scientific benchmark to evaluate the performance of relational database systems. TPC-X series study a series of benchmarks for testing and evaluation, including TPC-C and TPC-E for online transaction procession (OLTP), TPC-H and TPC-DS for online analytical processing (OLAP), etc. CH-Benchmark [17] mixes the TPC-C and TPC-H to measure one database's capability for both OLTP and OLAP. Set query benchmark [16] chooses a list of "basic" set queries from the aspects of document search, direct marketing, and decision support to evaluate the performance of one database. Star Schema Benchmark [18] employs a star schema to measure the performance of database products in support of classical data warehousing applications.

However, all existing benchmarks for RDBMS are not suitable for IMDB, due to insufficient consideration of the main characteristics of IMDB. In [19,20], some researchers try to explore the performance of IMDB. However, they only use response time and throughput as metrics without testing and evaluating other critical properties of IMDBs, such as compression ratio and cache miss. In contrast, our benchmark proposed a novel schema, metrics and workloads which are especially designed for IMDB, as illustrated in the introduction part.

## 3  Metrics

The MemTest metrics are computed from the information collected during the workload run. It can be divided into three groups: basic performance measures, CPU measures and memory measures.

The **basic performance measure** is response time. It includes the total time of twelve queries and two transactions. This metric is common in both IMDBs and disk-based databases. Since IMDBs store their data in main physical memory and employ different optimizations to structure and organize data, we design CPU measures and memory measures especially for the IMDBs.

The **CPU Measures** include CPU Usage and Cache Miss. CPU Usage records the CPU utilization during the data processing. After executing the workload in the order of 12 queries and 2 transactions, the CPU usage can be computed by the equation: $CPU\ usage = \frac{the\ Total\ CPU\ time - the\ CPU\ Waiting\ time}{the\ Total\ CPU\ time} * 100\%$. The Total CPU Time records the time interval of process from its beginning to the end. The CPU Waiting Time records the idle time of CPU during process. Since different hardwares have different architectures of CPU, it is hard to devise an universal equation to represent cache miss. In this study, we employ the interface provided by the hardware or the existing performance tool such as *perf* and *oprofile* in linux to get cache miss.

In the existing works, many technologies are proposed to improve CPU and cache efficiency, such as Partially Decomposed Storage Model (PDSM), Decomposed Storage Mode (DSM) and JiT Processing technology [21]. A good IMDB should employ advanced storage and processing technology to improve the CPU and cache at the same time.

The **Memory Measures** include Compression Rate and Minimal Memory Space.

– Compression Ratio (CR). There existed many compression techniques for IMDBs, such as dictionary compression, domain encoding and run-length encoding [22]. Different IMDBs may employ different compression algorithms. In this study, compression ratio (CR@SF) is used to reflect the compression capability of an IMDB. As mentioned above, an IMDB system tends to store compressed data in the memory to save memory cost. An excellent IMDB system is capable of loading more data into memory. Let $S_{Disk}$ denote the disk space allocation, and $S_{Mem}$ the memory occupation. CR@SF is defined as: $CR@SF = \frac{S_{Mem}}{S_{Disk}} * 100\%$.

– Minimal Memory Space (MMS). This metric describes the minimal memory space to execute the workload efficiently under a given SF value. In general, an IMDB system requires large amounts of memory to run the workload efficiently. In other words, almost all of the workloads can be conducted without significant I/O cost. It is worth noting that MMS is not equal to the size of disk data after compression, since it also needs to consider the memory allocation to execute the workloads. In fact, some IMDB systems may fail to work due to the memory deficiency.

## 4 Data Model

In this section, we devise a data model based on the inter-bank transaction applications.

### 4.1 Table Structure

Our database schema contains one big table with more than 200 columns and five small tables, as shown in Fig. 1. Due to the limitation of space, we only list common columns in the big table and then introduce the important attributes in all table. The **TRANSACTION_DETAIL** table details the transaction information, including the receive institution, the forward institution, the card, the merchant, the term, and so on. In fact, the table has more than 200 columns. Due to the space limitation, we use the COLUMN1 to COLUMN200 to represent the columns that cannot be listed in the table. The **RESP_INFO** table uses RESP_CD to judge whether the transaction is successful. Specifically, there are many kinds of response codes, among which the success code are unique and there are many failure codes since there exist many reasons for a failure transaction. The attributes RESP_NAME, RESP_DESC and RESP_TYPE illustrate a response code more detailedly. The attribute VALID_STATE record whether the code is currently valid. The **BRANCH_INFO** table stores the information of branches, including its name, nation, city, street and so on. The **MCHNT_INFO** table stores the information of merchants, including name, type of a merchant, address and so on. The **INSTITUTION_INFO** table records the information of institutions. It is worth noting that there are two types of institutions: receive institution (identified by rcv_ins_id) to process the transaction of a customer and forward institution (identified by fwd_ins_id) to

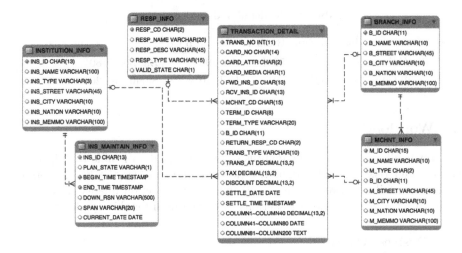

**Fig. 1.** The database schema of MemTest

forward a transaction apply of a customer. All the institutions are stored in this table. The **INS_MAINTAIN_INFO** table stores the abnormal information of institutions. It include the begin time and end time of an event, the reason for this event, current date, and so on.

## 4.2 Data Generation and Distribution

In this study, we implement a lightweight data generation in Java and it is fully platform independent. In fact, we also collect a real dataset from an inter-bank company. However, the dataset are not allowed to be published due to commercial reason. Table 2 lists the detailed distribution of data after analyzing a realistic applications. Sepecificly, 60 % of all institutions (ins), 70 % of all merchants (Mchnt) and 50 % of all branches are located in big cities. 90 % of transaction are successful (Suc) according to Response Codes (Resp_Cd). In all transaction, big institutions (Big Ins) occupies 80 % (Trans Freq). In order to simulate real scenarios, we devise a simple method to support the uniform and skew distribution of data in inter-bank transaction applications. For the attributes conforming uniform distribution, we employ $RND$ function to randomly generate data. For the attributes conforming skewed distribution, we first learn a parameter from the real data distribution as the skewed rate and then generate a dataset conforming the skew distribution. For instance, given a list with five items $\{r_1, r_2, ..., r_5\}$, from the real data set we learn that the item $r_1$ appears more frequently and its skewed rate is 0.7, then we use $RND$ function to generate a number $n$ randomly. If $1 \leqslant n \leqslant 7$, $r_1$ will be selected to populate the field, otherwise a item in $\{r_2, ..., r_5\}$ will be selected randomly. This generator also provides scalable volumes of raw data based on a scale factor (SF).

**Table 2.** The detail of data distribution

|  | Ins | Mchnt | Branch |  | Resp_Cd |  | Trans Freq |
|---|---|---|---|---|---|---|---|
| Big Cities | 60 % | 70 % | 50 % | Suc | 90 % | Big Ins | 80 % |
| Small Cities | 40 % | 30 % | 50 % | Fail | 10 % | Small Ins | 20 % |

## 5  Workload

In inter-bank transaction, the analysis of transaction data is able to help institutions find risky and abnormal transactions and uncover the key factors of transaction failure solve problems and improve transaction quality. Hence, we design a set of queries and transactions to be executed for this application.

### 5.1  Business Cases

In this section, we identify four groups of queries including twelve queries and two transactions to simulate realistic business cases in inter-bank applications.

**Query Group 1. Institutions' Transaction Statistics and Analysis.** In this group, three queries are designed to help find the institutions which have a lower success rate. Consequently, the decision maker can improve the transaction especially for the institutions which have larger transaction amount and frequency. Specifically, query 1.1 computes the total amount, average tax and average discount of each institution for each day, query 1.2 computes the transaction frequency of each institution for each day, query 1.3 keeps count of the successful rate and failure rate of each institution for each day according to the response code.

**Query Group 2. Transaction Quality Analysis.** Queries in this group are targeted at discovering the transaction quality through analyzing the behavior of response codes. Query 2.1 finds the total number and amount of transactions for each day according to different response codes. In order to cast more concentration on the failure transactions, query 2.2 keeps count of the total number of failure transactions according to the failure response code. Thereafter, among the failure transactions, query 2.3 computes the total number of institutions, terms and merchants.

**Query Group 3. Transaction Compliance Analysis.** In the realistic inter-bank transaction applications, besides the transaction quality, another important type of transaction is compliance transaction. It means that the transaction is successful but illegal, such as testing transactions for a system. Query 3.1 computes low-amount transaction number and average transaction amount according to the term, merchant and branch, query 3.2 record the sign-in transactions and return the top 10 merchants and terms in the branch, query 3.3 return the high-rate failure transactions for each day.

**Query Group 4. Institutions' Abnormal Statistics and Analysis.** This group mainly is designed to detect institutions' abnormal information, caused by the hardware or software in an institution. First, query 4.1 divides the institutions into three classes (e.g. high-incident, middle-incident, low-incident) according to their abnormal frequency in the history. Then, query 4.2 computes the number of cards, terms, branches and merchants, etc. for each abnormal institution in the transaction. Query 4.3 computes the total amount, average tax and average discount of transactions during the institutions incident.

**Transaction 1. Generating New Transactions.** Every day, there are huge data which will enter the database system in the inter-bank application. For instance, China Unionpay can generate at least 30,000,000 transactions. Given this situation, we design this transaction to insert data into table **TRANS-ACTION_DETAIL**. Other business cases in this part include the statistics of transaction number every day.

**Transaction 2. Capturing the Most Abnormal Institutions.** This transaction is targeted at finding the most abnormal institution. Different from query group 4, this transaction will compute the number of abnormal events in each institution and find the most abnormal institution. Then, the detailed transactions of this institution will be analyzed.

**List 1.** The specification of Q2.3

```
SELECT RESP_CD, RESP_NAME, INS_NAME, TERM_ID, TERM_TYPE
    , M_NAME, COUNT(*) TRANS_NUM
FROM TRANSACTION_DETAIL T, RESP_INFO R, INSTITUTION_INFO
    I, MCHNT_INFO M
WHERE T.RETURN_RESP_CD=R.RESP_CD AND T.RCV_INS_ID=I.
    INS_ID AND T.MCHNT_CD=M.M_ID AND RESP_TYPE='DELTA1'
    AND VALID_STATE='DELTA2' AND SETTLE_DATE BETWEEN '
    DELTA3' AND 'DELTA4'
GROUP BY RESP_CD, RESP_NAME, INS_NAME, TERM_ID, TERM_TYPE,
    M_NAME
ORDER BY TRANS_NUM DESC;
```

## 5.2  Technical Details

As mentioned before, IMDBs have redesigned the store organization and processing algorithms. For instance, during the query processing, most conventional database systems attempt to minimize disk access, whereas IMDBs focus more on processing costs. In our workload, we design complex queries to cover different operators in database field. Specifically, all queries have aggregations, 90 % of queries have join operators and a transaction includes update operations. In the workload, a query may have some arguments, like **SETTLE_DATE**, **VALID_STATE** and **RESP_TYPE**. We list the specification of query 2.3 in List 1 as an example. All the arguments can be randomly replaced by a query generator from a predefined dictionary. Optionally, we also offer a fixed replace strategy and then generate the workload to simulate daily work in inter-bank applications.

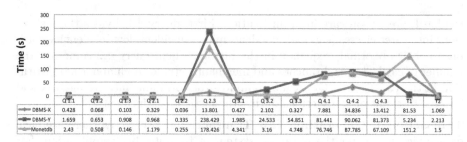

| | Q1.1 | Q1.2 | Q1.3 | Q2.1 | Q2.2 | Q2.3 | Q3.1 | Q3.2 | Q3.3 | Q4.1 | Q4.2 | Q4.3 | T1 | T2 |
|---|---|---|---|---|---|---|---|---|---|---|---|---|---|---|
| DBMS-X | 0.428 | 0.068 | 0.103 | 0.329 | 0.036 | 13.801 | 0.427 | 2.102 | 0.327 | 7.881 | 34.836 | 13.412 | 81.53 | 1.069 |
| DBMS-Y | 1.659 | 0.653 | 0.908 | 0.968 | 0.335 | 238.429 | 1.985 | 24.533 | 54.851 | 81.441 | 90.062 | 81.373 | 5.234 | 2.213 |
| Monetdb | 2.43 | 0.508 | 0.146 | 1.179 | 0.255 | 178.426 | 4.341 | 3.16 | 4.748 | 76.746 | 87.785 | 67.109 | 151.2 | 1.5 |

**Fig. 2.** Testing for query and transaction (SF=2)

# 6    Experiments

In this section, we conduct a series of experiments to benchmark three different IMDBs: DBMS-X, DBMS-Y, and Monetdb. DBMS-X is an IMDB system which supports features like column-based storage and queries, data compression and parallel processing. DBMS-Y is based on row-based storage with features like durability, query optimization, recoverability etc. Monetdb is an open-source column-oriented database with features like data compression and query optimizers. The server under test has 0.5 TB memory and 8 CPUs. Specifically, it has three levels cache: L1 cache (private) size is 32 KB, L2 cache (private) size is 256 KB and L3 cache (shared within a CPU) size is 30 MB. By default, the memory size that three IMDBs can use is 90 % of total memory and all queries are generated previously and keep unchanged during the benchmark run.

**Testing for Response Time.** First, we evaluate the response time of all queries and transactions, as shown in Fig. 2. It is observed that execution time of Q1.1, Q1.2, Q1.3, Q2.1, Q2.2, Q3.1 and T2 are low in three IMDBs since their operators are relatively simple. We can see that Q2.3 costs the most of time in three IMDBs. It is because multiple joins need to be processed in this query, as shown in List 1. In summary, for all queries, DBMS-Y has the longest execution time and Monetdb since it is based on the row-oriented store and have to load attributes which are not required to process, resulting to the increment of processing cost. In Fig 2, it is also shown that DBMS-Y performs better than DBMS-X and Monetdb when transaction one is executed. This is an interesting implication. Actually, most of IMDBs based on column-oriented store (e.g. DBMS-X) will not directly modify the physics of the most original records when they are required to be updated. These systems only mark the original records useless and store new records in another area. However, through this technology, there also still exists performance difference between column-oriented store and row-oriented store systems since the former need to maintain extra structures to record and maintain the data to be updated.

We also test the performance of IMDBs when multiple users execute the whole workload. In this test, we implement a parallel tool to simulate the operator of multiple users. From Table 3, we can see that the results of three IMDBs

**Table 3.** Testing in the multi-user model

| Users<br>IMDB | 2 | 4 | 6 | ... |
|---|---|---|---|---|
| DBMS-X | 558.12s | 1412.51s | ⩾ 1500s | ... |
| DBMS-Y | 824s | ⩾ 1800s | – | ... |
| Monetdb | 613s | ⩾ 1500s | – | ... |

are similar and DBMS-X has a slightly better scalability than other two IMDBs. For three IMDBs, we do not list all the results due to the long execution time.

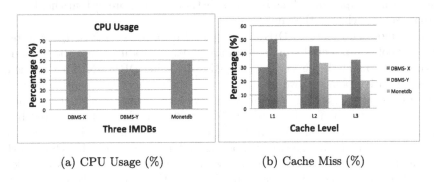

(a) CPU Usage (%)                 (b) Cache Miss (%)

**Fig. 3.** Testing for CPU usage and cache miss (SF=2)

**Testing for CPU Measures.** In this section, we test CPU usage and cache miss to know more about IMDBs under test. We employ the one-user model. In Fig. 3(a), we can see that three IMDBs have a relatively high CPU usage (>40 %) and DBMS-X and Monetdb behave better than DBMS-Y. This phenomenon can be explained based on the physical organization. If the tuples are stored in the row-store style, some attributes which are not needed must be loaded into the cache. In the same size of cache, it will increase the percentage of useless data, resulting to the frequent data exchange between cache and memory. Thereafter, the CPU waiting time will increase, resulting to the reduction of CPU usage. However, if the data is organized based on the column-store style, For the same size of cache, it can load more data tuples that are referenced. This will reduce the data exchange between cache and memory. The CPU usage will also increase. In Fig. 3(b), we continued the evaluation of cache miss to verify the difference for different organization styles. It has a corresponding result to the CPU usage.

**Testing for Memory Measures.** In the following part, Fig. 4(a) describes the compression scalability with the increment of SF. It is observed that DBMS-X and monetdb have better compression ratios, approximately 0.1 and 0.3. DBMS-Y

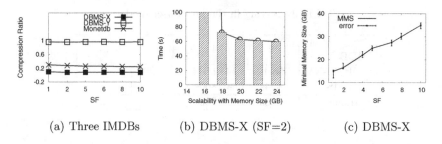

(a) Three IMDBs         (b) DBMS-X (SF=2)         (c) DBMS-X

**Fig. 4.** Testing and evaluation for *Compression Ratio* and *Minimal Memory Size*

almost has no compression. That is because in the column-store system a field stores the data sets which have the same type.

Since more and more IMDBs are employed to process analytic tasks in real time and column-oriented store system are widely introduced into the industry, we continue to test the effectiveness of MMS by using DBMS-X. This test is in fact an iterative plan and we need to change the memory size continuously. In other words, after assigning the SF value, we also need to assign the memory space in each iteration and record the behaviors. The data set generated by a given SF value is common for all iterations. During each iteration, we restart the database server, change the memory space for use, execute the whole work-load, and observe the change of response time. In this way, MMS is recorded. Figure 4(b) shows that when the memory size is less than 18 GB, the memory processing capability decreases a lot. It is shown that with the increment of memory space, the current memory space is sufficient to process data if the execution time increment is slow. Otherwise, the execution time will grow up significantly, or even the system will crash down. Thereafter, Fig. 4(c) lists the trend of MMS with the increment of SF.

In this section, we have tried our best to guarantee the objectiveness of experimental results. The conclusion is that column-oriented store system is good at complex analysis and row-oriented store system can deal with transaction processing well in IMDBs. In summary, as a novel data processing technology IMDBs have gained huge performance improvement.

# 7    Conclusion

In this study, we proposed an IMDB benchmark, named MemTest. This benchmark takes special consideration of main characteristics of IMDBs. Accordingly, novel metrics are especially designed for testing and evaluating IMDBs. In MemTest, we provide a schema based on inter-bank transaction applications. The workload is devised to cover OLAP and OLTP operations. Finally, experiments are conducted to verify the effectiveness and efficiency of our benchmark by implementing and running it on three systems. For future work, we may devise more complex queries and conduct the experiments on more IMDBs.

**Acknowledgement.** Our research is supported by the 973 program of China (No. 2012CB316203), NSFC (No. 61370101), Shanghai Knowledge Service Platform Project (No. ZF1213), Innovation Program of Shanghai Municipal Education Commission (14ZZ045) and the Natural Science Foundation of Shanghai (14ZR1 412600).

# References

1. Fedorova, A., Seltzer, M.I., Small, C., Nussbaum, D.: Performance of multithreaded chip multiprocessors and implications for operating system design. In: USENIX Annual Technical Conference, General Track, pp. 395–398 (2005)
2. Färber, F., Cha, S.K., Primsch, J., Bornhövd, C., Sigg, S., Lehner, W.: Sap hana database: data management for modern business applications. In: SIGMOD Record, pp. 45–51 (2011)
3. Lahiri, T., Neimat, M.A., Folkman, S.: Oracle timesten: an in-memory database for enterprise applications. IEEE Data Eng. Bull. **36**, 6–13 (2013)
4. Diaconu, C., Freedman, C., Ismert, E., Larson, P.A., Mittal, P., Stonecipher, R., Verma, N., Zwilling, M.: Hekaton: sql server's memory-optimized oltp engine. In: SIGMOD Conference, pp. 1243–1254 (2013)
5. HyperSQL: Hypersql. http://hsqldb.org
6. SQLite: Sqlite. http://www.sqlite.org/mostdeployed.html
7. MemSql: Memsql. http://www.memsql.com
8. Monetdb: Monetdb. https://www.monetdb.org/Home
9. Carey, M.J., DeWitt, D.J., Naughton, J.F.: The oo7 benchmark. In: Buneman, P., Jajodia, S. (eds.): SIGMOD Conference, pp. 12–21. ACM Press (1993)
10. Carey, M.J., DeWitt, D.J., Naughton, J.F., Asgarian, M., Brown, P., Gehrke, J., Shah, D.: The bucky object-relational benchmark (experience paper). In: SIGMOD Conference, pp. 135–146 (1997)
11. Schmidt, A., Waas, F., Kersten, M.L., Carey, M.J., Manolescu, I., Busse, R.: Xmark: a benchmark for xml data management. In: VLDB, pp. 974–985 (2002)
12. Carey, M.J., Ling, L., Nicola, M., Shao, L.: EXRT: towards a simple benchmark for XML readiness testing. In: Nambiar, R., Poess, M. (eds.) TPCTC 2010. LNCS, vol. 6417, pp. 93–109. Springer, Heidelberg (2011)
13. Cooper, B.F., Silberstein, A., Tam, E., Ramakrishnan, R., Sears, R.: Benchmarking cloud serving systems with ycsb. In: SoCC, pp. 143–154 (2010)
14. Ghazal, A., Rabl, T., Hu, M., Raab, F., Poess, M., Crolotte, A., Jacobsen, H.A.: Bigbench: towards an industry standard benchmark for big data analytics. In: SIGMOD Conference, 1197–1208 (2013)
15. Gupta, M.K., Verma, V., Verma, M.S.: In-memory database systems - a paradigm shift, pp. 333–336 (2014)
16. Gray, J.: Benchmark Handbook: For Database and Transaction Processing Systems. Morgan Kaufmann Publishers Inc., San Francisco (1992)
17. Cole, R., Funke, F., Giakoumakis, L., Guy, W., Kemper, A., Krompass, S., Kuno, H.A., et al.: The mixed workload ch-benchmark. In: DBTest, pp. 8:1–8:6 (2011)
18. Rabl, T., Poess, M., Jacobsen, H.A., O'Neil, P.E., O'Neil, E.J.: Variations of the star schema benchmark to test the effects of data skew on query performance. In: ICPE, pp. 361–372 (2013)
19. Tözün, P., Pandis, I., Kaynak, C., Jevdjic, D., Ailamaki, A.: From a to e: analyzing tpc's oltp benchmarks: the obsolete, the ubiquitous, the unexplored. In: EDBT, pp. 17–28 (2013)

20. Liu, D., Luan, H., Wang, S., Qin, B.: Main memory database TPC-H workload characterization on modern process. J. Softw. **19**, 2573–2584 (2008)
21. Pirk, H., Funke, F., Grund, M., Neumann, T., Leser, U., Manegold, S., Kemper, A., Kersten, M.L.: Cpu and cache efficient management of memory-resident databases. In: ICDE, pp. 14–25 (2013)
22. Müller, I., Ratsch, C., Frber, F.: Adaptive string dictionary compression in in-memory column-store database systems. In: EDBT, pp. 283–294 (2014)

# DSIMBench: A Benchmark for Microarray Data Using R

Shicai Wang, Ioannis Pandis, Ibrahim Emam, David Johnson, Florian Guitton,
Axel Oehmichen, and Yike Guo[✉]

Data Science Institute, William Penney Laboratory, Imperial College London,
South Kensington Campus, London SW7 2AZ, UK
{s.wang11,i.pandis,i.emam,david.johnson,f.guitton,
axelfrancois.oehmichen11,y.guo}@imperial.ac.uk
http://dsg.doc.ic.ac.uk

**Abstract.** Parallel computing in R has been widely used to analyse microarray data. We have seen various applications using various data distribution and calculation approaches. Newer data storage systems, such as MySQL Cluster and HBase, have been proposed for R data storage; while the parallel computation frameworks, including MPI and MapReduce, have been applied to R computation. Thus, it is difficult to understand the whole analysis workflows for which the tool kits are suited for a specific environment. In this paper we propose DSIMBench, a benchmark containing two classic microarray analysis functions with eight different parallel R workflows, and evaluate the benchmark in the IC Cloud testbed platform.

**Keywords:** Benchmark · R · MPI · MapReduce

## 1 Introduction

Data mining techniques applied to microarray data convert raw intensity values into useful information. R is one of the most popular data mining software tools used for medical research. With masses of data accumulating from translational research studies involving high-throughput sequencing, many high performance databases, such as MySQL Cluster [1], PostgreSQL Cluster [2], MongoDB [3] and HBase [4], and parallel computing frameworks, including Message Passing Interface (MPI) [5] and MapReduce [6], are being integrated into the traditional microarray analysis tool, R [7]. Though these new methods greatly improve the performance of R, they greatly complicate the whole analysis workflow. For example, all the databases and parallel frameworks mentioned above form eight different R workflows. Many datasets are required to fully evaluate the performance of each workflow. Thus, hundreds of, or even thousands of, tests must be performed in order to robustly evaluate and determine the most efficient and effective workflow.

Our motivation is to find an effective big data solution for our open source knowledge management software platform tranSMART [8], which was originally

© Springer International Publishing Switzerland 2014
J. Zhan et al. (Eds.): BPOE 2014, LNCS 8807, pp. 47–56, 2014.
DOI: 10.1007/978-3-319-13021-7_4

developed by Johnson&Johnson for in-house clinical trial and knowledge management requirements in translational studies. For the needs of various collaborative translational research projects, an instance of tranSMART is hosted at Imperial College London and has been configured to use an Oracle relational database for back-end storage. It currently holds over 70 million gene expression records. When querying the database simultaneously for hundreds of patient gene expression records, a typical exercise in translational studies, the record retrieval time can currently take up to several minutes. Furthermore, some typical analyses using R, such as marker selection and data clustering, can take up to several minutes, or even hours. These kinds of response times impede applications performed by researchers using this deployed configuration of tranSMART. Anticipating the requirement to store and analyse next generation sequencing data, where the volume of data being produced will be in the TB or PB range, the current performance exhibited by tranSMART is unacceptably poor.

In this paper, we present DSIMBench (Data Science Institute Microarray Benchmarks), which uses two common translational medical applications with six representative data mining workflows, and evaluate the benchmark on the IC Cloud [9] testbed.

## 2    Related Work

Benchmarks play a significant role in all domains. SPEC [10] benchmarks are gold standards used by many processor manufacturers and researchers to measure the effectiveness of their inventions. Popular benchmarking suites designed for specific application domains are also well accepted, such as TPC-H [11] for database systems, SPLASH [12] for parallel architectures, and MediaBench II [13] for media and communication processors.

Many bioinformatics benchmark suites are widely in use, such as BioBench [14], BioPerf [15] and MineBench [16]. These benchmarks contain several applications in common, including BLAST, FASTA, Clustalw, and Hmmer. The bioinformatics applications presented in DSIMBench differ from those included in these benchmark suites. BioBench contains only serial workloads. Bioperf only uses a few parallelized applications. Even in MineBench which contains full-fledged OpenMP parallelized codes of all bioinformatics work-loads, no large-scale computing framework has been integrated, such as MPI and MapReduce. In contrast to the above benchmarks, DSIMBench focuses on R scalability and performance for big data technologies with microarray data.

## 3    R with High Performance Plugins

### 3.1    Data Distribution

A standard vanilla R workflow loads the entire data before performing calculations. However, R provides many interfaces to different kinds of storage systems such as built-in functions (e.g. CSV reader), for local file system access (e.g. Linux ext4), DBI plugins for relational database access (e.g. MySQL Cluster), and RHadoop plugin for interfacing with key-value database clusters (e.g. HBase).

## 3.2  Parallel Computation

There are many high performance R plugins that parallelize calculationd for CPU cores within one machine or for CPU cores across machines configured in a computing cluster. MPI and MapReduce are two representative technologies used for big data. For MPI, the R Snowfall [17] plugin is a usability wrapper around the Rmpi [18] plugin for more usable development of parallel R programs. Rmpi is a widely used MPI interface for the Local Area Multicomputer (LAM) [19], with MPICH2 [20], a MPI implementation. For MapReduce, the RHadoop plugin is a representative interface for the Apache Hadoop ecosystem [21], including Hadoop Distributed File System (HDFS), MapReduce and HBase.

## 4  DSIMBench Workflows

We designed eight R workflows based on different data distributions and computational solutions, as shown in Table 1. The first three workflows (W1–W3) are created to test the data loading performance. Each workflow loads data from one of three data sources, including local file system ext4, relational database MySQL Cluster and key-value database HBase, and performs computation in vanilla R. Workflow W4 acts as a baseline test for the parallel computations. Workflows W5 and W6 test only the performance on the parallelization of the computation in R, as the data is delegated directly from the master node through direct network sockets. Finally workflows W7 and W8 test both the data loading and parallel computation in combination, where W7 loads data to the worker nodes using the fastest data loading workflow chosen from test results of W1–W3 with MPI, while W8 loads data using RHBase and computes using MapReduce.

**Table 1.** The DSIMBench workflows.

| Workflows | Data loading | Computation | Data source | Parallel method |
|-----------|--------------|-------------|-------------|-----------------|
| W1 | Single process | N/A | ext4 | N/A |
| W2 | Single process | N/A | HBase | N/A |
| W3 | Single process | N/A | MySQL Cluster | N/A |
| W4 | N/A | Single process | N/A | Vanilla R |
| W5 | N/A | Multiple cores | N/A | MPI |
| W6 | N/A | Multiple cores | N/A | MapReduce |
| W7 | Multiple processes | Multiple cores | Best DB | MPI |
| W8 | Multiple processes | Multiple cores | RHBase | MapReduce |

### 4.1 Data Source Performance Assessment

In workflows W1–W3, as shown in Fig. 1, third-party R plugins are used to connect to the respective data sources. Hence, it is possible that the different implementations of these R plugins could interfere with the performance of the data source. In order to better assess the performance of the data sources in workflows W1–W3 we performed data loading tests using a user-based high-level API written in Java to directly load data from each data source to identify how much of the performance is affected by that middleware layer in R. The fastest source is then tested via a R plugin. If this R plugin on the source outperforms the other data sources via Java APIs, this data source will be used in the following W4–W6 tests.

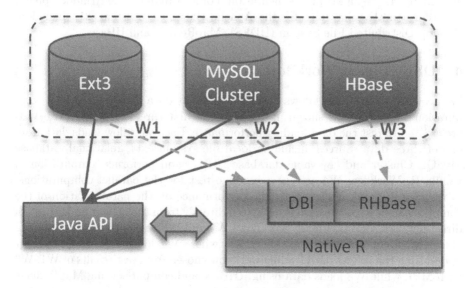

**Fig. 1.** Diagram illustrating how the loading test is organised.

### 4.2 Parallel Computation Benchmark Workflows

W4 in Fig. 2 is introduced as the baseline. W5 shows R Snowfall MPI computation via a LAM/MPICH2 cluster. The data distribution consists of two sequential steps: data loading and data copy. The input data matrix is loaded into LAM master node and then fully copied to all MPICH2 slave nodes. The calculation is carried out by the Snowfall `sfLapply()` function. `sfLapply()` mediates the distributed calculation in the slave nodes and collects the results. W6 indicates RHadoop MapReduce computation via a Apache Hadoop cluster. MapReduce in W6 utilises HDFS as Mapper task data source. Thus, the data distribution consists of two sequential steps. First, the input data matrix is split into data blocks and then uploaded into HDFS. The number of data blocks depend on the number of Mappers. After a MapReduce computation, all the results are stored in HDFS.

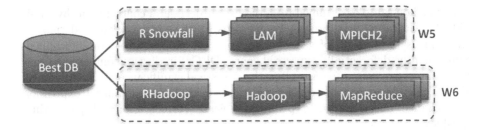

**Fig. 2.** Diagram illustrating how the parallel framework test is organised.

W7 in Fig. 3 manipulates MPICH2 tasks to load directly from the fastest data source based on W1–W3. If ext4 is applied to W7, the matrix data file will be split into data blocks and copied to each worker during the data preparation. The number of data blocks depends on the MPICH2 number. W8 manipulates Mapper tasks to load directly from HBase to test the built-in MapReduce HBase performance. RHadoop launches Mapper tasks without data loading. Each Mapper task loads data via built-in access to HBase Scanner and computes concurrently.

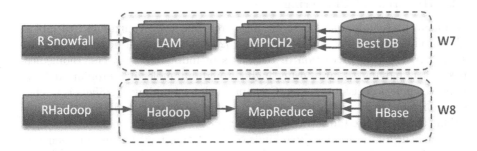

**Fig. 3.** Diagram illustrating R MapReduce with HBase.

## 5  DSIMBench Applications

### 5.1  Marker Selection

High-throughput gene expression analysis is a technique used to uncover disease specific gene signatures and gain further insight into disease mechanisms. In the past decade, gene expression measurements have shifted from quantitative assays capable of measuring the expression of single genes, to assays capable of assessing the levels of the majority of expressed genes in cells, tissues or organisms of interest. DNA microarray chips are the common technology platform used in recent years and are capable of simultaneous determination the entire human "transcriptome". In complex disease research, including diseases such as asthma and chronic obstructive pulmonary disease (COPD), microarray experiments are performed on samples obtained from disease subjects and control (healthy) individuals. After the initial pre-processing steps which reduce background "noise", the expression intensity of genes present on each chip/sample are determined. Subsequently, deferentially expressed genes (DEGs) in disease compared to control samples are computed as well as the statistical significance of the difference. Finally, DEGs can be filtered by the relative levels of differential expression (fold-change) and significance (p-values; typically corrected for multiple testing: q-values).

The basic use case is to create a cohort between the patients and the control. For some more complicated ones, many clinical measurements are utilised to generate cohorts. A test case below was carried out using a large publicly available transcriptomic dataset taken from NCBI GEO [22] concerning Multiple Myeloma (GEO accession GSE24080; Popovici et al., 2010 [23]). The dataset contains 559 subjects' gene expression data produced by an Affymetrix GeneChip Human Genome U133 Plus 2.0 Array. The cohorts are generated depending on patient medical therapies and survival time. This test case is utilised in workflows W1–W3 to test the data query in different number of subjects based on different cohorts.

## 5.2 Hierarchical Clustering

Genomic, proteomic and metabolic measurements have contributed to molecular profiling based patient stratification [24], such as identification of disease subgroups and the prediction of responses of individual subjects. Biomedical research is moving towards using high-throughput molecular profiling data to improve clinical decision-making. One approach for building classifiers is to classify subjects based on their molecular profiles. Unsupervised clustering algorithms can be utilised for stratification purposes.

Our benchmark applies three kinds of correlation methods used to generate correlation matrices that are used by the hierarchical clustering algorithm in tranSMART - the Pearson product-moment correlation, Spearmans rank-order correlation, and Euclidean distance correlation. The test case below was carried out using a large publicly available transcriptomic dataset taken from NCBI GEO concerning leukemia (GEO accession GSE13204; Kohlmann et al., 2014 [25]). The dataset contains 2325 subjects' gene expression data produced by an Affymetrix GeneChip Human Genome U133 Plus 2.0 Array. The correlation matrix calculations could either be implemented on Hadoop, a popular and well supported distributed data storage and computation framework that supports MapReduce, or be implemented for distributed execution in R using Snowfall, a parallel computing package for R scripts. In this benchmark, all W4–W6 and the fastest one in W1–W3 are utilised to test the hierarchical clustering method.

## 6    Results

We performed the data loading test on marker selection and parallel tests on the hierarchical clustering on 4 virtual machines in our IC Cloud implementation.

The 4 VMs works on two physical machines with each 24 core and 64 GB memory. Each physical machine hosts 2 VMs.

- Operating system: CentOS Linux 2.6.18-308.24.1.el5xen
- CPU: 144 cores (Intel(R) Xeon(R) CPU E5-2630 0 @ 2.60GHz)
- Memory: 384 GB, DDR3, 1066 MHz
- Disk array: 24 TB (Huawei, OceanStor S5500T)
- Virtual machine: 8 virtual CPU cores, 8 GB memory

### 6.1   Data Loading Test

We chose 5 cohorts (10 cases) of different data size and the whole dataset of GSE24080 for a marker selection exercise, shown in Fig. 4. In the Java API test, loading from ext4 file system outperforms all the other data sources. A widely used vanilla R function scan greatly utilised for R the CSV file reading. RHBase performs better than scan only in the first data size. In the following parallel R test we choose ext4 as the data source.

### 6.2   R Parallel Framework Test

We utilised R function `rdist()` in package fields to calculate euclidean distance matrices, function `cor()` to calculate the Pearson and Spearman correlation matrices and function `hclust()` to cluster the correlation matrices. The result of W4 and W5 to compare parallel frameworks, shown in Fig. 5(a), indicates that when using the smaller MULTIMYEL dataset, MPI and MapReduce perform slower than vanilla R. W5 suffers from slow data transmission. The result of W7 and W8 to compare multi-thread data losing using different data sources, shown in Fig. 5(b), indicates it is faster for MPICH2 to load data directly from ext4 than HBase. W8 suffers from the long time RHBase data loading, as shown in Fig. 4. The vanilla R computation (W2) performs best in the small dataset, but does not scale up well in the large dataset. In the large dataset the better parallel methods in Fig. 5(a) (W6) and (b) (W7) are utilised to compare to W4. W6 and W7 outperformed W4. Though W6 computation time is a little longer than W7, W6 outperform W7 due to the faster data preparation.

## 7   Discussion

As shown in Fig. 5, the parallel methods suffer from data communication overheads such as transferring data to each worker, worker management and collecting data from the workers post-computation. But when size of the dataset increases, the advantages of parallel methods overcome these overheads. In Fig. 5(c), W6 and W7 have similar computation times, but W6 benefits from faster data preparation using HDFS. We considered using RHadoop with HBase at the beginning, however RHBase demonstrates poor data loading performance and is consequently much slower than the HBase Java API. RHBase does not perform well due to

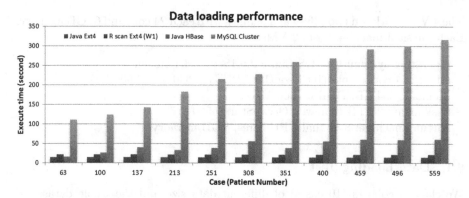

**Fig. 4.** Bar chart showing the performance evaluation in our data loading tests.

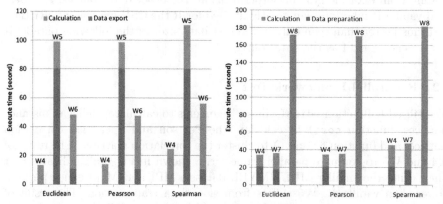

(a) Parallel R using single thread loading in small dataset.

(b) Parallel R using multi-thread loading in small dataset.

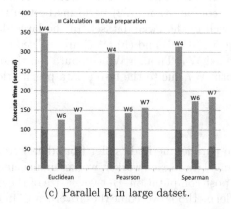

(c) Parallel R in large datset.

**Fig. 5.** Bar chart showing the performance evaluation in our computation tests.

the slow HBase Thrift [26] server. If RHBase could be implemented via rJava and HBase Java API, it may perform much better. Also, the data loading tests should introduce concurrent data loading tests before computation tests and full tests. As shown in Fig. 5, though parallel approaches can improve the data loading, optimisation of the matrix computation should not be neglected. R matrix calculations use a pure array object to gain significant performance using the CPU cache. Parallel methods divide a big matrix into small pieces and executes calculations by the low-speed R loop functions that cannot be pre-loaded in CPU cache due to potential R branch sentences. This is the reason why parallel methods can only perform 2 or 3 times faster than vanilla R when 32 CPU cores are utilised.

## 8   Conclusion

Big microarray data analysis using R is gaining significant focus as it's data access and computationally intensive workloads are in dire need to optimise their performance. We believe a new data mining benchmark is required to thoroughly analyse these analysis workflows and propose the most optimal workflow setup for them. In this paper, we presented DSIMBench, a benchmark containing two classic microarray analysis functions with six different parallel R workflows, and evaluated the benchmark in IC Cloud testbed platform.

**Acknowledgment.** This research was partially supported by the Innovative R&D Team Support Program of Guangdong Province (NO. 201001D0104726115), China, Johnson & Johnson Pharmaceutical and Research Comp, and Innovative Medicines Initiative (IMI), EU Grant Code 115446.

## References

1. MySQL Cluster CGE. http://www.mysql.com/products/cluster/
2. Momjian, B.: PostgreSQL: introduction and concepts. J. Digit. Imaging Off. J. Soc. Comput. Appl. Radiol. **22**, 462 (2001). doi:10.1007/s10278-007-9097-5
3. Dirolf, K.C., Dorif, M.: MongoDB: The Definitive Guide. O'Reily Media, Sebastopol (2011)
4. George, L.: HBase The Definitive Guide. O'Reily Media, Sebastopol (2008)
5. Anon: MPI: A message passing interface. In: Proceedings of the Supercomputing Conference, pp. 878–883 (1993). doi:10.1109/SUPERC.1993.1263546
6. Dean, J., Ghemawat, S.: MapReduce: simplified data processing on large clusters. Commun. ACM **51**(1), 107–113 (2008). doi:10.1145/1327452.1327492
7. R Development Core Team: R: a language and environment for statistical computing. R Foundation for Statistical Computing, Vienna, Austria (2008). ISBN 3-900051-07-0, http://www.R-project.org
8. Athey, B.D., Braxenthaler, M., Haas, M., Guo, Y.: tranSMART: an open source and community-driven informatics and data sharing platform for clinical and translational research. In: Proceedings of the AMIA Joint Summits on Translational Science 2013, pp. 6–8, PMCID: PMC3814495 (2013)

9. Guo, L., Guo, Y., Tia, X.: IC cloud: a design space for composable cloud computing. In: Proceedings - 2010 IEEE 3rd International Conference on Cloud Computing, CLOUD 2010, pp. 394–401 (2010). doi:10.1109/CLOUD.2010.18

10. Henning, J.L.: SPEC CPU2006 benchmark descriptions. ACM SIGARCH Comput. Archit. News **34**(4), 1–17 (2006). doi:10.1145/1186736.1186737

11. TPC-H Benchmark. http://www.tpc.org/tpch/

12. Woo, S.C., Ohara, M., Torrie, E., Singh, J.P., Gupta, A.: The SPLASH-2 programs: characterization and methodological considerations. In: Proceedings 22nd Annual International Symposium on Computer Architecture (1995). doi:10.1109/ISCA.1995.524546

13. Fritts, J.E., Steiling, F.W., Tucek, J.A.: MediaBench II Video: Expediting the Next Generation of Video Systems Research. In: Proceedings of the SPIE, Embedded Processors for Multimedia and Communications II, vol. 5683, pp. 79–93 (2005)

14. Albayraktaroglu, K., Jaleel, A., Wu, X., Franklin, M., Jacob, B., Tseng, C.W., Yeung, D.: BioBench: a benchmark suite of bioinformatics applications. In: ISPASS 2005 - IEEE International Symposium on Performance Analysis of Systems and Software, vol. 2005, pp. 2–9 (2005). doi:10.1109/ISPASS.2005.1430554

15. Bader, D.A., Li, Y., Li, T., Sachdeva, V.: BioPerf: a benchmark suite to evaluate high-performance computer architecture on bioinformatics applications. In: Proceedings of the 2005 IEEE International Symposium on Workload Characterization, IISWC-2005, vol. 2005, pp. 163–173 (2005). doi:10.1109/IISWC.2005.1526013

16. Narayanan, R., Ozisikyilmaz, B., Zambreno, J., Memik, G., Choudhary, A.: MineBench: a benchmark suite for data mining workloads. In: Proceedings of the 2006 IEEE International Symposium on Workload Characterization, IISWC - 2006, pp. 182–188 (2006). doi:10.1109/IISWC.2006.302743

17. Knaus, J., Porzelius, C., Binder, H.: Easier parallel computing in R with snowfall and sfCluster. Source **1**, 54–59 (2009)

18. Yu, H.: Rmpi: parallel statistical computing in R. R News 2, 10–14 (2002). http://cran.r-project.org/doc/Rnews/Rnews_2002-2.pdf

19. Squyres, J.M.: A component architecture for LAM/MPI. ACM SIGPLAN Not. (2003). doi:10.1145/966049.781510

20. Bridges, P., Doss, N., Gropp, W., Karrels, E., Lusk, E., Skjellum, A.: User Guide to MPICH, a Portable Implementation of MPI. Argonne National Laboratory, 9700, 60439–64801 (1995)

21. White, T.: Hadoop: The Definitive Guide. O'Reilly Media, Sebastopol (2012)

22. Barrett, T., Wilhite, S.E., Ledoux, P., Evangelista, C., Kim, I.F., Tomashevsky, M., Soboleva, A.: NCBI GEO: Archive for functional genomics data sets - Update. Nucleic Acids Res.,**41** (2013). doi:10.1093/nar/gks1193

23. Popovici, V., Chen, W., Gallas, B.G., Hatzis, C., et al.: Effect of training-sample size and classification difficulty on the accuracy of genomic predictors. Breast Cancer Res. **12**(1), R5 (2010)

24. Stoughton, R.B., Friend, S.H.: How molecular profiling could revolutionize drug discovery. Nat. Rev. Drug Discov. **4**, 345–350 (2005). doi:10.1038/nrd1696

25. Kohlmann, A., Kipps, T.J., Rassenti, L.Z., Downing, J.R., et al.: An international standardization programme towards the application of gene expression profiling in routine leukaemia diagnostics: the Microarray Innovations in LEukemia study prephase. Br. J. Haematol. **142**(5), 802–807 (2008)

26. Apache Thrift. http://thrift.apache.org

# A Benchmark to Evaluate Mobile Video Upload to Cloud Infrastructures

Afsin Akdogan[✉], Hien To, Seon Ho Kim, and Cyrus Shahabi

Integrated Media Systems Center, University of Southern California,
Los Angeles, CA, USA
{aakdogan, hto, seonkim, shahabi}@usc.edu

**Abstract.** The number of mobile devices (e.g., smartphones, tablets, wearable devices) is rapidly growing. In line with this trend, a massive amount of mobile videos with metadata (e.g., geospatial properties), which are captured using the sensors available on these devices, are being collected. Clearly, a computing infrastructure is needed to store and manage this ever-growing large-scale video dataset with its structured data. Meanwhile, cloud computing service providers such as Amazon, Google and Microsoft allow users to lease computing resources with varying combinations of computing resources such as disk, network and CPU capacities. To effectively use these emerging cloud platforms in support of mobile video applications, the application workflow and resources required at each stage must be clearly defined. In this paper, we deploy a mobile video application (dubbed *MediaQ*), which manages a large amount of user-generated mobile videos, to Amazon EC2. We define a typical video upload workflow consisting of three phases: (1) video transmission and archival, (2) metadata insertion to database, and (3) video transcoding. While this workflow has a heterogeneous load profile, we introduce a single metric, frames-per-second, for video upload benchmarking and evaluation purposes on various cloud server types. This single metric enables us to quantitatively compare main system resources (disk, CPU, and network) with each other towards selecting the right server types on cloud infrastructure for this workflow.

**Keywords:** Mobile video systems · Spatial databases · Cloud computing · Big video data · Benchmarking

## 1 Introduction

With the recent advances in video technologies and mobile devices (e.g., smartphones, tablets, wearable devices), massive amounts of user generated mobile videos are being collected and stored. According to Cisco's forecast [7], there will be over 10 billion mobile devices by 2018 and 54 % of them will be *smart* devices, up from 21 % in 2013. Accordingly, mobile video will increase 14-fold between 2013 and 2018, accounting for 69 % of total mobile data traffic by the end of the forecasted period. Clearly, this vast amount of data brings a major scalability problem in any computing infrastructure. On the other hand, cloud computing provides flexible resource arrangements that can instantaneously scale up and down to accommodate varying workloads. It is projected that the total economic impact of cloud technology could be

© Springer International Publishing Switzerland 2014
J. Zhan et al. (Eds.): BPOE 2014, LNCS 8807, pp. 57–70, 2014.
DOI: 10.1007/978-3-319-13021-7_5

$1.7 trillion to $6.2 trillion annually in 2025 [8]. Thus, the large IT service providers such as Amazon, Google, and Microsoft, are ramping up cloud infrastructures.

One key question is how to evaluate the performance of mobile video applications on these cloud infrastructures and select the appropriate set of resources for a given application. Suppose a mobile user wants to upload a video to a cloud server along with its metadata (e.g., geospatial properties of video such as camera location and viewing direction), which are captured and extracted using the sensors embedded on the mobile devices. Note that this kind of geospatial metadata enables advanced data management, especially in very large-scale video repositories. For example, the performance of a spatial query such as a range query, which can find all video frames that overlap with a user-specified region [2], can be significantly enhanced using spatial metadata. When we upload captured videos with metadata from mobile device to cloud, this upload operation consists of three stages which require different computing system resources: (1) *network* to transfer videos from mobile clients to the cloud servers (i.e., network bandwidth), (2) *database* to insert metadata about the uploaded videos (i.e., database transaction), and (3) *video transcoding* to change the resolution of uploaded videos to use less storage and bandwidth (i.e., CPU processing power). These phases are executed in sequence; therefore, inefficiency in any step slows down the performance of video applications. A benchmark to evaluate such an application needs to identify the system resources used at each stage, compare them with one another *quantitatively* and spot which resource(s) becomes the *bottleneck* in the workflow of the application. Once the bottlenecks are detected, the servers with the right specifications can be selected and configured accordingly on cloud.

There exists a challenge in evaluating the performance of a large scale video application on cloud because most of the benchmarking studies in the cloud computing context focus on evaluating either the performance of Big Data processing frameworks such as Hadoop and Hive [25, 26] or NoSQL data-stores rather than considering all system resources a mobile video application requires. In particular, some benchmarks are designed for social networking applications [17], online transaction processing (OLTP) [9, 10, 19] and simple key-value based put-get operations [16, 18]. These benchmarks only emphasize the impact of the database system on the overall performance. In addition, a recent study measures the impact of virtualization on the networking performance in the data centers [6]. However, this study only measures packet delays and TCP/UDP throughput, and packet loss among virtual machines.

In this paper, we define a single (cross-resource) metric to evaluate the uploading workflow of video applications on cloud and present an end-to-end benchmark. In particular, we use a throughput, the number of *processed frames per second*, as the metric and compare the performance of system resources (e.g., network, disk, CPU) with one another. To this extent, we deployed one exemplary mobile video application called MediaQ, which we developed on the Amazon EC2 platform, and conducted extensive experiments on various server types. Specifically, we used the smallest and the largest instance at each server group (e.g., *disk-optimized, CPU-optimized, general-purpose*) to identify the *lower and upper performance bound*. Our experimental results show that CPU drastically slows down the entire system and becomes the bottleneck in the overall performance. Our experiments also show that simply selecting high-end CPU-optimized servers does not resolve the problem entirely. Therefore, we propose two techniques to enhance the CPU throughput: (1) reducing video quality and (2) enabling multithreading. Our study serves as the first step towards understanding the end-to-end performance characteristics of cloud resources in terms of resource-demanding video applications.

The remainder of this paper is organized as follows. Section 2 provides the necessary background. The benchmark design and experimental results are presented in Sects. 3 and 4, respectively. Related work is discussed in Sect. 5. Subsequently Sect. 6 concludes the paper with the directions for future work.

## 2 Background

Before we present our results and findings, we briefly introduce a typical example of resource intensive mobile video application (MediaQ) and available server types in the data centers to prepare for the rest of the discussion.

### 2.1 MediaQ: Mobile Multimedia Management System

MediaQ [2, 3] is an online media management framework that includes functions to collect, organize, search, and share user-generated mobile videos using automatically tagged geospatial metadata. MediaQ consists of a MediaQ server and a mobile app for smartphones and tablets using iOS and Android. User-generated-videos (UGV) can be uploaded to the MediaQ server from users' smartphones and they are then displayed accurately on a map interface according to their automatically collected geo-tags and other metadata information such as the recorded real time, camera location, and the specific direction the camera was pointing. Media content can be collected in a casual or on-demand manner. Logged in participants can post specific content requests that will automatically generate an alert with other participants who are near a desired content assignment location to entice them to record using their phones.

The schematic design of the MediaQ system is summarized in Fig. 1. Client-side components are for user interaction, i.e., the Mobile App and the Web App. The Mobile App is mainly for video capturing with sensed metadata and their uploading. The Web App allows searching the videos and issuing spatial crowdsourcing task requests to collect specific videos. Server-side components consist of Web Services, Video Processing, GeoCrowd Engine, Query Processing, Account Management, and Data Store. The Web Service is the interface between client-side and server-side components. The Video Processing component performs transcoding of uploaded videos so that they can be served in various players. At the same time, uploaded videos are analyzed by the visual analytics module to extract extra information about their content such as the number of people in a scene. We can plug in open source visual analytics algorithms here to achieve more advanced analyses such as face recognition among a small group of people such as a user's family or friends. Automatic keyword tagging is also performed at this stage in parallel to reduce the latency delay at the server. Metadata (captured sensor data, extracted keywords, and results from visual analytics) are stored separately from uploaded media content within the Data Store.

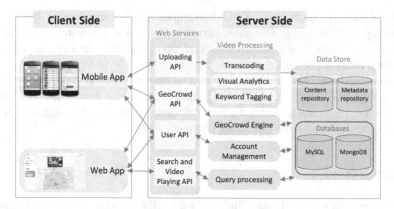

**Fig. 1.** Overall structure of MediaQ system

Query Processing supports effective searching for video content using the metadata in the database. Finally, task management for spatial crowdsourcing can be performed via the GeoCrowd engine.

## 2.2 Cloud Server Type Classification

Recently the computing resources on cloud have become more granular since service providers use virtualization techniques to manage physical servers and provide a wide selection of server types optimized to fit different use cases [1]. These types comprise varying combinations of CPU, memory, storage, and networking capacity and give users the flexibility to choose an appropriate combination of resources. Specifically, server types are clustered into six groups where each group consists of several options with varying computational capabilities. Table 1 depicts a classification of the server groups presently offered by the biggest three service providers along with the prices (dollars/hour) of the smallest and the largest server at each group. As shown, the pricing varies widely across

**Table 1.** Categorization of the server types with the prices (dollars/hour) of the smallest and the largest servers at each group.

| Type | Amazon EC2 | | Microsoft Azure | | Google Compute | |
|------|------------|--|------------------|--|-----------------|--|
| | Price ($/hour) | | Price ($/hour) | | Price ($/hour) | |
| | Smallest | Largest | Smallest | Largest | Smallest | Largest |
| General purpose ($m$) | 0.07 | 0.56 | 0.02 | 0.72 | 0.077 | 1.232 |
| Compute optimized ($c$) | 0.105 | 1.68 | 2.45 | 4.9 | 0.096 | 0.768 |
| Memory optimized ($r$) | 0.175 | 2.8 | 0.33 | 1.32 | 0.18 | 1.44 |
| Disk optimized ($i$) | 0.853 | 6.82 | – | – | – | – |
| Micro ($t$) | 0.02 | 0.044 | – | – | 0.014 | 0.0385 |
| GPU | 0.65 | 0.65 | – | – | – | – |

the server types within each service provider. For example, the most expensive machine in Microsoft Azure is *245* times more costly than the cheapest one (*4.9/0.02*). This ratio is *97* for Amazon EC2 and *102* for Google Compute Engine, respectively. Clearly, such a huge discrepancy across the server types makes the selection of appropriate set of resources critical in hosting an application on these cloud platforms.

# 3  Benchmark Design

In this section, we first explain our measurement methodology and then discuss the metric we used in the experimental evaluation.

## 3.1  Methodology

There are three main components in the performance evaluation of large scale mobile video systems such as MediaQ which requires different system resources: (1) **network** to transfer videos from mobile clients to a cloud server (i.e., network bandwidth), (2) **database** to insert metadata about the uploaded videos (i.e., database transaction), and (3) **video transcoding** to change the resolution of uploaded videos which is a common operation in video services (i.e., CPU processing power). Specifically, we measure the *upload performance* which involves these three phases that are executed in sequence. Upon recording a video, mobile clients retrieve metadata (i.e., GPS signals, field of views, etc.) from the video. Subsequently, along with the video data, they upload the metadata in JSON format to the server. Once a video is uploaded, the metadata is inserted into the database and the video is transcoded, which is required to either support different formats (e.g., MP4, WAV) or to reduce video quality due to limited network bandwidth when being displayed later. Therefore, the videos are not retrievable until transcoding task is completed, and hence overhead in any component can degrade the overall performance of video applications. Our goal is to define a single metric, examine these components individually using this metric, and detect which phase slows down the system. To this extent, we deployed MediaQ server side code on the EC2 servers running a video upload service implemented in PHP. The service can receive multiple video files simultaneously. We then run multiple clients which transfer large amount of videos concurrently using the upload service.

## 3.2  Metric

We introduce a single metric, *processed-frames-per-second*, to evaluate the performance of three main components. For network performance, we straightforwardly report the number of transferred frames per second. For database performance, we report the number of frames inserted per second. Note that we do not insert the video data but its spatio-temporal metadata to the database. The metadata are collected at video capturing time by mobile devices and transferred to the server, thus the database cost is only composed of inserting a set of metadata (i.e., per frame) from memory into database. Similar to the standalone version of MediaQ, we selected MySQL database

**Table 2.** Hardware specifications of the smallest and largest servers of 4 server types on EC2.

| Type | Memory | CPU | Disk | Network bandwidth |
|---|---|---|---|---|
| *m-small* | 3.75 GB | 1 VCPU | 4 GB SSD | No info. |
| *c-small* | 3.75 GB | 2 VCPUs | 32 GB SSD | No info. |
| *r-small* | 15.25 GB | 2 VCPUs | 32 GB SSD | No info. |
| *i-small* | 30.5 GB | 4 vCPUs | 800 GB SSD | No info. |
| *m-large* | 30 GB | 8 VCPU | 160 GB SSD | No info. |
| *c- large* | 60 GB | 32 VCPUs | 640 GB SSD | No info. |
| *r- large* | 244 GB | 32 VCPUs | 2 × 320 GB SSD | 10 Gigabit Ethernet |
| *i- large* | 244 GB | 32 vCPUs | 8 × 800 GB SSD | 10 Gigabit Ethernet |

installed on EC2 servers. For transcoding performance, we use the number of trans-coded frames per second. Once a video arrives at the server, MediaQ transcodes it using FFMPEG [13], which is a widely used video solution. In order to measure the maximum throughput, we perform a stress test on the cloud server by generating a large amount of real videos and uploading them to the server simultaneously and continuously for a significant amount of time.

## 4 Performance Evaluation

In this section, we first present an overall cost analysis of the three components in the workflow and show how the server types impact the performance. Subsequently, we evaluate transcoding and database components in more detail, and finally present performance-cost results.

### 4.1 Overall Cost Analysis

In this set of experiments, we used the smallest servers on Amazon EC2 in four instance families: *general purpose (m)*, *compute-optimized (c)*, *memory-optimized (r)* *and disk-optimized (i)* and measured the throughputs (See Table 2 for hardware specifications of Amazon EC2). To fully utilize multi-core CPUs available at the servers, we enabled multi-threading on database and transcoding parts. Specifically, we first run the experiments using one thread ($T = 1$), and then increase the number of threads $T$ by one to run the experiment again. The point where throughput cannot be improved further is the maximum throughput that the server can achieve. Note that there is no index built on the metadata table in the database and we take advantage of bulk insert, where 1,000 rows are written into disk as one transaction which reduces the disk I/O significantly. For transcoding tasks, we reduce the video resolution from 960 × 540 to 480 × 270.

Figure 2a illustrates the throughput comparison where a single large video with 24 fps (frame per second) was uploaded to the server. We observed that other than general purpose instance, the performance difference between the optimized servers ($c, r, i$) is not significant even though the prices vary widely such that *i-small* is *8* times

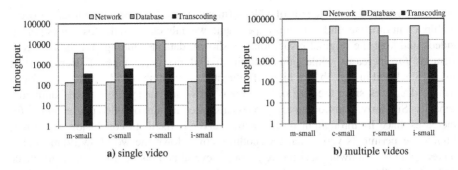

**Fig. 2.** Comparison of system components on smallest servers of 4 server types (log-scale).

more expensive than *c-small*. As shown, database can handle almost *two* and *three orders of magnitude* more frames than network and transcoding, respectively.

Fully utilizing the network bandwidth, real systems can handle concurrent video uploads to the server; therefore, in the next experiments, we did a stress test where multiple videos were uploaded simultaneously until the network bandwidth was saturated. As illustrated in Fig. 2b, network throughput increases significantly; however, database and transcoding remain almost constant. This is because we already enabled bulk insert and multi-threading to ensure the maximum performance even in the case of a single video upload. Another observation is that *transcoding*, which shows the lowest throughput, becomes a major *bottleneck* in the workflow.

The frames per second (*fps*) value in video recording has a direct impact on the performance in our experiments. However, database throughput is independent of *fps* in our target application MediaQ. This is because, *fps* value ranges from 15 to 120 in new generation smartphone cameras; however, regardless of *fps*, we select one metadata per each second using a sampling technique [3] and store it for all the frames in the corresponding second. This is a real-world phenomenon since metadata includes geospatial attributes such as the camera location and viewing direction which do not change significantly within a second. This approach widens the gap between the throughput of database and those of other components even further. Figure 3a depicts a comparison of the system resources under various *fps* values on a c-small instance,

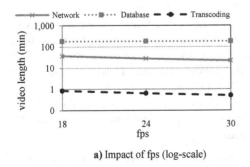

| fps | Network | Database | Transcoding |
|-----|---------|----------|-------------|
| 24  | - 25.1% | %0       | - 23.3%     |
| 30  | - 39.8% | %0       | - 38.3%     |

**a)** Impact of fps (log-scale)          **b)** Decrease over fps=18

**Fig. 3.** The video length (min) that each component can process in a second for various *fps*.

where the metric is total length of videos (minute) that each component can process within a unit time (second). For this experiment, we only changed the fps values [14] of videos and kept the original resolution. As shown, database throughput remains constant as *fps* increases while others diminish and the percentage of decrease over fps = 18 is listed in Fig. 3b. For higher frame rates, the length of video that network can handle decreases since the size of the videos grow and saturate the fixed bandwidth capacity. Similarly, transcoding can process a shorter amount of video per second as *fps* increases since its throughput on a specific server is fixed. In conclusion, these preliminary experiments verify that transcoding slows down the workflow dramatically; therefore, in the following section we propose several approaches to enhance this piece and measure the impact of each proposed technique.

## 4.2    Transcoding Performance

It is crucial to improve the transcoding performance because newly uploaded videos are not retrievable for use until their transcoding tasks are completed. That is the main reason behind the delay between the uploading and viewing time in many video-based applications. There are two ways to make transcoding faster: (1) enabling multi-threading, and (2) reducing the size of the output file, which results in a lower video quality. We explain these two approaches in turn.

**Multi-threading.** One natural way to improve the performance is utilizing multi-core CPUs available in the servers and *scale-up*. There are two techniques to increase the throughput on cloud. First, running a multi-threaded *ffmpeg* process (*MT*) on a single video and decrease the total amount of time to transcode it. Second, running a single-threaded *ffmpeg* processes in parallel on multiple videos (*PST*). In the following set of experiments, we use the largest *compute-optimized* (c3.xlarge) server with 32 vCPU's.

Figure 4a illustrates the effect of varying number of threads while transcoding a video. In this specific experiment, we used a 230 MB video in *AVI* format as input and reduced the resolution from 960 × 540 to 480 × 270 in two different video output types, MP4 and AVI. As shown, the total time does not decrease linearly as the number of threads increases. As stated in Amdahl's Law [6], a parallel algorithm is as fast as its sequential, non-parallelizable portion which dominates the total execution time. For *ffmpeg*, after *4 threads* the performance gain becomes insignificant no matter how many CPUs are used, which verifies that ffmpeg does not scale up.

Another way to increase throughput is running single-thread *ffmpeg* processes in parallel where each thread handles a single video. Figure 4b depicts the throughput performance of these two techniques. Since *ffmpeg* does not scale well as the number of threads increases, the throughput remains almost constant. However, throughput increases almost linearly for *PST* until all 32 CPUs are fully utilized. That is because while *MT* technique suffers from low parallelism, *PST* can utilize available CPUs better. After the CPUs are saturated, the performance goes down for both *MT* and *PST* due to resource contention across the threads.

**Reducing Video Quality.** In this set of experiment, we investigate the impact of *resolution* and *type* of the outputted video on the performance. Table 3 presents the

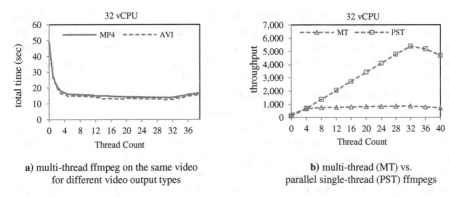

a) multi-thread ffmpeg on the same video          b) multi-thread (MT) vs.
for different video output types          parallel single-thread (PST) ffmpegs

**Fig. 4.** Scale up performance of transcoding.

transcoding throughputs for converting a video with dimensions 960 × 540 to smaller resolutions. The percentage of increase in the throughput is listed as well for clarity of the presentation. As expected, the result shows that throughput increases significantly as the resolution decreases. However, the percentage improvement diminishes when the output video resolution becomes too smaller (i.e., 60 × 32) because loading the input video, frame by frame, is a constant cost which largely contributes to the total transcoding cost. In addition, we also observe that the results are similar for different output formats (*mp4* and *avi*).

**Table 3.** Transcoding throughput for mp4 and avi types with various output resolutions. The input video is in.m4v format with 960 × 540 resolution.

| Output resolution | MP4 | | AVI | |
|---|---|---|---|---|
| | Throughput | % improvement | Throughput | % improvement |
| 480 × 270 | 623 | – | 626 | – |
| 240 × 136 | 842 | 35 % | 839 | 34 % |
| 120 × 68 | 980 | 57 % | 982 | 57 % |
| 60 × 34 | 1038 | 66 % | 1048 | 67 % |

### 4.3   Database Performance

In this section, we measure the database throughput on both the smallest and largest instances at each server group to show the lower and upper *performance bounds*. In addition, we investigate the impact of indexing on throughput and test if it changes the best performer server. Throughput is measured using iterative multi-threading approach. First, we run the experiment with a single thread and repeat the experiment increase the number of threads by one until no throughput improvement is observed. Then, we report the maximum throughput as the result.

Metadata information is stored in *video_metadata* database table which consists of *13* columns where average length of a row is *319* bytes. Figure 5a and b illustrate the effect of server types on the database throughput where the smallest and largest instances in general purpose (*m*), compute-optimized (*c*), memory-optimized (*r*) and disk-optimized (*i*) are clustered together. As shown in Fig. 5a, where there is no index on *video_metadata* table, the smallest disk-optimized server (*i*) slightly outperforms others. With the largest instances, compute-optimized (*c*) server provides slightly better performance than others. This is because while small servers contain 2 to 4 CPUs, large ones have 8 to 32 CPUs and compute-optimized machines might be better in managing *concurrent threads*. Note that, even though metadata insertion is an I/O-intensive task, disk-optimized machines do not expressively outperform other instances. The reason is that video data is mostly *append-only*, where the updates to the dataset are rare after the insertion. Disk-optimized instances are tuned to provide fast *random I/O*; however, in *append-only* datasets random access is not much used. Another observation is that optimized machines perform at least *2.5* times better than the general purpose one.

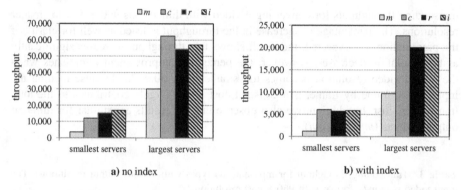

**a) no index**                    **b) with index**

**Fig. 5.** Database insertion throughput on the smallest and largest servers in 4 instance families.

**Effect of Indexing.** To investigate how indexing influences throughput at each server, we built 2 indices on *video_metadata* table. Specifically, a B-tree index on the time field and hash index on the keywords fields, which is a good indexing strategy that allows efficient range search over the time and effective equality search on the keywords associated with the videos. As depicted in Fig. 5b, for both smallest and largest server groups, *compute-optimized* instances show better performance unlike the no-index scenario. The reason is that indices are kept in memory and index update is a CPU intensive task. Another observation is that indexing degrades the performance considerably, where throughput approximately drops to *1/3* of the no-index scenario due to extra high index maintenance cost.

## 4.4 Performance-Cost Analysis

In this section, we discuss how the performance-to-price ratios of different server types. Figure 6 illustrates the number of frames per dollar that each component can process using the smallest servers in each server group. In this specific example, we uploaded

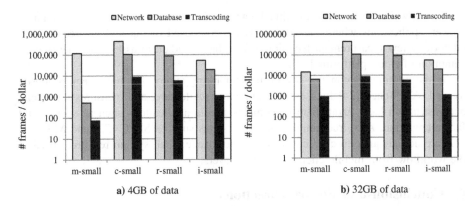

**Fig. 6.** Number of frames that can be processed for each dollar spent (log-scale).

multiple videos with 24 fps to the servers and enabled multithreading mode both on database insertion and transcoding. Note that performance-price ratio is sensitive to the total amount of data we upload since more nodes need to be employed when the storage capacity of a single server is exceeded. In Fig. 6a, we present the result for 4 GB of data which can fit all server types (See Table 2 for server specifications). As shown in Fig. 6a, compute-optimized server (*c-small*) outperforms other server types in all aspects. In addition, disk-optimized server (*i-small*) is not cost efficient for video uploads since there is not much need for random disk access as discussed in the previous section and the data size is small. In Fig. 6b, we present the results for 32 GB of data which 8 times exceeds the m-small node. In this scenario, we need allocate 8 small servers; therefore, cost efficiency dramatically drops for m-small server.

## 5   Related Work

With the increasing popularity of Big Data processing frameworks, several benchmarks have been proposed to evaluate various offline operations (e.g., grep, sort, aggregation, etc.) on popular frameworks such as Hadoop and Hive [25, 26]. Meanwhile, a number of benchmarks have been developed to measure the scalability of NoSQL and NewSQL databases. These benchmarks only emphasize the impact of the database system on the overall performance rather than considering all the resources an end-to-end mobile video application requires. In particular, some benchmarks are designed for social networking applications [17], online transaction processing (OLTP) [9, 10, 19] and simple key-value based put-get operations, which are heavily used in web applications [16, 18]. In addition, there are a few recent studies that measure the impact of virtualization on the networking performance in the data centers [6, 20]. However, these studies only measure packet delays and TCP/UDP throughput, and packet loss among virtual machines. Similar to our approach, CloudCmp [21] measures the performance of elastic computing and persistent storage services offered by cloud service providers. However, CloudCmp separates computing and storage instances, and employs different metrics for performance evaluation and cross-platform comparisons.

Moreover, while we focus on multimedia applications CloudCmp covers a wide range of web applications where the workloads are composed of put and get requests. In addition, a few measurement techniques have been studied to assess the energy consumption of cloud platforms [12].

In the multimedia context, several benchmarking approaches have been proposed as well. ALPBench [24] focuses on multi-core architecture, and measures the thread and instruction-level parallelism of complex media applications such as speech and face recognition. Also traditional benchmark suites such as SPEC [22] and MiBench [21] are not adequate to characterize the performance of all system resources used in the workflow from mobile clients to cloud servers.

## 6 Conclusion and Future Directions

In this paper, we proposed a single frame-based metric which can measure the performance of three main system components on cloud infrastructure for a large-scale mobile video application, especially for uploading videos from mobile clients to cloud servers. To this extent, we first deployed our mobile video management system, MediaQ, to Amazon EC2, separated video upload workflow into three phases and identified the system resources used at each stage. Using our metric, we spotted CPU as the main *bottleneck* that slows down the entire system performance. Subsequently, we proposed several approaches to enhance CPU throughput and concluded that running multiple single-threaded transcoding processes increases throughput linearly with the number of CPUs. In addition, to benchmark various server types available on EC2, we conducted our experiments on four different server families, specifically, on the smallest and the largest instance of servers to identify the *lower* and *upper performance bound*. Our experimental results show that compute-optimized machines provide the best performance for a resource intensive mobile video application.

We believe that our approach will help users to make more informed decisions in choosing server types while deploying mobile video applications to cloud infrastructures. In addition, such a cross-resource metric can be used to calculate performance-to-price ratios. As a next step, we plan to extend our frame-based metric to measure: (1) mobile devices' computing and storage capabilities, and (2) other server side processes such as query processing (e.g., range query). Moreover, we also would like to partition our dataset and scale out to multiple servers.

**Acknowledgements.** This research has been funded in part by NSF grants IIS-1115153 and IIS-1320149, the USC Integrated Media Systems Center (IMSC), and unrestricted cash gifts from Google, Northrop Grumman, Microsoft, and Oracle. Any opinions, findings, and conclusions or recommendations expressed in this material are those of the author(s) and do not necessarily reflect the views of any of the sponsors such as the National Science Foundation.

## References

1. Amazon EC2. http://docs.aws.amazon.com/AWSEC2/latest/UserGuide/instance-types.html
2. MediaQ Framework. http://mediaq.usc.edu

3. Kim S.H., Lu Y., Constantinou, G., Shahabi, C, Wang, G, Zimmermann, R.: MediaQ: mobile multimedia management system. In:5th ACM Multimedia Systems Conference, pp. 224–235. ACM, New York (2014)

4. Oracle. http://docs.oracle.com/cd/B12037_01/appdev.101/b10795/adfns_in.htm

5. Wang, G., Eugene, T.S.: The impact of virtualization on network performance of amazon EC2 data center. In: 29th Conference on Information Communications (INFOCOM), pp. 1163–1171. IEEE Press, Piscataway (2010)

6. Amdahl G.: Validity of the single processor approach to achieving large-scale computing capabilities. In: Spring Joint Conference (AFIPS), pp. 483–485. ACM, New York (1967)

7. Cisco's Forecast. http://www.cisco.com/c/en/us/solutions/collateral/service-provider/ip-ngn-ip-next-generation-network/white_paper_c11-481360.pdf

8. Mc Kinsey's Forecast. http://www.mckinsey.com/insights/business_technology/disruptive_technologies

9. Curino, C., Difallah, D.E., Pavlo, A., Cudre-Mauroux, P.: Benchmarking OLTP/Web databases in the cloud: the OLTP-bench framework. In: 4th International Workshop on Cloud Data Management, pp. 17–20. ACM, New York (2012)

10. Kossmann, D., Kraska, T., Loesing, S.: An evaluation of alternative architectures for transaction processing in the cloud. In: International Conference on Management of Data (SIGMOD), pp. 579–590. ACM, New York (2010)

11. TPC: TPC-W 1.8. TPC Council (2002)

12. Cuzzocrea, A., Kittl, C., Simos, D.E., Weippl, E., Xu, L. (eds.): CD-ARES 2013. LNCS, vol. 8127, pp. 272–288. Springer, Heidelberg (2013)

13. Ffmpeg Library. www.ffmpeg.org

14. Android. http://developer.android.com/reference/android/hardware/Camera.Parameters.html#setPreviewFpsRange

15. Venkata, S., Ahn, I., Jeon, D., Gupta, A., Louie, C., Garcia, S., Belongie, S., Taylor, M.: Sd-vbs: The San Diego vision benchmark suite. In: International Symposium on Workload Characterization (IISWC), pp. 55–64. IEEE, Washington, DC (2009)

16. Cooper, B.F., Silberstein, A., Tam, E., Ramakrishnan, R., Sears, R.: Benchmarking cloud serving systems with YCSB. In: 1st ACM Symposium on Cloud Computing (SoCC), pp. 143–154. ACM, New York (2010)

17. Barahmand, S, Ghandeharizadeh, S.: BG: a benchmark to evaluate interactive social networking actions. In: Sixth Biennial Conference on Innovative Data Systems Research (CIDR), Asilomar, CA, USA (2013)

18. Patil, S., Polte, M., Ren, K, Tantisiriroj, W., Xiao, L., López, J, Gibson, G, Fuchs, A., Rinaldi, B.: YCSB++: benchmarking and performance debugging advanced features in scalable table stores. In: 2nd ACM Symposium on Cloud Computing (SOCC). ACM, New York (2011)

19. Gray, J.: The Benchmarking Handbook for Database and Transactions Systems. Morgan Kaufman, San Francisco (1992)

20. Ballani, H., Costa, P., Karagiannis, T., Rowstron, A.: Towards predictable datacenter networks. In: 17th International Conference on Data Communications (SIGCOMM), pp. 242–253. ACM, New York (2011)

21. Li, A., Yang, X., Kandula, S., Zhang, M.: CloudCmp: comparing public cloud providers. In: 10th International SIGCOMM Conference on Internet Measurements, pp. 1–14. ACM, New York (2010)

22. The Standard Performance Evaluation Corporation (SPEC). www.specbench.org

23. Guthaus, M., Ringenberg, J., Ernst, D., Austin, T., Mudge, T., Brown, R.: Mibench: a free, commercially representative embedded benchmark suite. In: International Symposium on Workload Characterization, pp. 3–14

24. Li, M.L., Sasanka, R., Adve, S.V., Chen, Y.K., Debes, E.: The ALPBench benchmark suite for complex multimedia applications. In: International Symposium on Workload Characterization, pp. 34–45. IEEE, Washington, DC (2005)
25. Luo, C., Zhan, J., Jia, Z., Wang, L., Lu, G., Zhang, L., Xu, C.Z., Sun, N.: CloudRank-D: benchmarking and ranking cloud computing systems for data processing applications. J. Front. Comput. Sci. 6(4), 347–362 (2012)
26. Wang, L., Zhan, J., Luo, C., Zhu, Y., Yang, Q., He, Y., Gao, W., Jia, Z., Shi, Y., Zhang, S., Zheng, C., Lu, G., Zhan, K., Li, X., Qiu, B.: BigDataBenchd: a big data benchmark suite from internet services. In: 20th IEEE International Symposium on High Performance Computer Architecture, pp. 488–499, Orlando, Florida, USA (2014)

# Benchmarking Replication and Consistency Strategies in Cloud Serving Databases: HBase and Cassandra

Huajin Wang[1,2], Jianhui Li[1(✉)], Haiming Zhang[1],
and Yuanchun Zhou[1]

[1] Computer Network Information Center,
Chinese Academy of Sciences, Beijing, China
{wanghj,lijh,hai,zyc}@cnic.cn
[2] University of the Chinese Academy of Sciences, Beijing, China

**Abstract.** Databases serving OLTP operations generated by cloud applications have been widely researched and deployed nowadays. Such cloud serving databases like BigTable, HBase, Cassandra, Azure and many others are designed to handle a large number of concurrent requests performed on the cloud end. Such systems can elastically scale out to thousands of commodity hardware by using a shared nothing distributed architecture. This implies a strong need of data replication to guarantee service availability and data access performance. Data replication can improve system availability by redirecting operations against failed data blocks to their replicas and improve performance by rebalancing load across multiple replicas. However, according to the PA-CELC model, as soon as a distributed database replicates data, another tradeoff between consistency and latency arises. This tradeoff motivates us to figure out how the latency changes when we adjust the replication factor and the consistency level. The replication factor determines how many replicas a data block should maintain, and the consistency level specifies how to deal with read and write requests performed on replicas. We use YCSB to conduct several benchmarking efforts to do this job. We report benchmark results for two widely used systems: HBase and Cassandra.

**Keywords:** Database · Replication · Consistency · Benchmark · Hbase · Cassandra · YCSB

## 1 Introduction

In recent years, it has become a trend to adopt cloud computing in the IT industry. This trend is driven by the rapid development of internet-based services such as social network, online shopping and web search engines. These cloud based systems need to deal with terabytes and even larger amounts of data, as well as keep the cloud service high reliable and available for millions of users. Such scenarios require cloud serving databases to be able to handle huge number of concurrent transaction timely (availability) and to increase their computing capacity during running time (scalability and elasticity). It is difficult for traditional databases to do such jobs, which motivates the

© Springer International Publishing Switzerland 2014
J. Zhan et al. (Eds.): BPOE 2014, LNCS 8807, pp. 71–82, 2014.
DOI: 10.1007/978-3-319-13021-7_6

development of new cloud serving databases to satisfy the above requirements. BigTable [1], developed by Google to support cloud data serving in the context of Big Data era, is the first of such databases, and it has inspired a variety of similar systems such as HBase [2] and Cassandra [3]. These systems are usually based on key-value stores, adopt a shared nothing distributed framework and therefore can scale out to thousands of commodity hardware at running time.

However, these new type of databases encounter the CAP [4] tradeoff: According to the CAP theorem, it is impossible for a distributed computer system to simultaneously provide all the three following guarantees: availability, consistency and partition tolerance. In practice, cloud serving databases are usually deployed on clusters consisting of thousands of commodity machines. In such clusters, network partitions are unavoidable: Failures of commodity hardware are very common and the update operations can barely be simultaneously performed on different nodes too, which mimics partitions. For such reasons, cloud serving databases must make tradeoff between availability and consistency. In order to keep high availability, most cloud serving databases choose a weaker consistency mechanism than the ACID transactions in traditional databases.

Also due to the partition problem, cloud serving databases usually use data replication to guarantee service availability and data access performance. Data replication can improve the system availability by redirecting operations against failed data blocks to their replicas, and improve data access performance by rebalancing load across multiple replicas. However, according to the PACELC [5] model, as soon as a distributed database replicates data, another tradeoff between consistency and latency arises. This tradeoff motivates us to figure out how the latency changes when we adjust the replication factor and the consistency level:

- *Replication factor:* The replication factor determine how many replicas a data block should maintain in a specific scenario? Replicas can be used for failover and to spread the load to them, which may imply that the higher replication factor, the better load balancing and the shorter request latency. However, such an assumption is questionable before we do some performance comparisons by changing the replication factor of the same cluster. Besides, the storage capacity are not unlimited. When we use a replication factor of $n$, the actual space occupied by the database is $n$ times the size of the records it originally intends to store. So, we should carefully make decisions on the replication factor.
- *Consistency level:* How to process read and write requests performed on replicas? For example, writes are synchronously written to all of the replicas in HBase to keep all replicas up to date, which may lead to high write latency. Asynchronously writing brings lower latency, however, in which replicas may be outdate.

We need to benchmark these tradeoffs to give answers to the above questions. In this work, we elaborate the benchmark methodology and show some results of this benchmarking effort. We report the performance results for two databases: HBase and Cassandra. We focus on the changes in request latency and throughput when the strategy of replication and consistency changes.

The paper is organized as follows. Section 2 provides a brief introduction to the strategy designs on replication and consistency in HBase and Cassandra. Section 3 discusses the methodology behind this benchmarking effort. Details about the benchmark and the testbed are also included in this section. Benchmark results are illustrated with corresponding analyses in Sect. 4. Section 5 reviews several related work on benchmarking efforts in this filed. We look to the future work in Sect. 6 and make conclusions in Sect. 7.

## 2 Database Design

This section investigates how each of HBase and Cassandra has been designed on data replication and how they try to keep consistency between replicas.

HBase provides strong consistency for both read and write. The clients cannot read inconsistent records until the inconsistency is fixed [6]. HBase doesn't write updates to disk instantly, instead, it saves updates in a write-ahead-log (WAL) stored in hard drive and then does in-memory data replication across different nodes, which increases the write throughput. In-memory files are flushed into HDFS when the size of them reaches the upper limit. HBase uses HDFS to configure the replication factor and save replicas.[1] Apparently, HBase prefer consistency to availability when it makes the CAP tradeoff.

Unlike HBase, Cassandra supports a tunable consistency level. There are three well known consistency levels in Cassandra: level ONE, level ALL and level QUORUM. Literally, the names represent the number of replicas on which the read/write must succeed before response to the client. The consistency level for read and write can be set separately in Cassandra. Reasonable choices on consistency are listed below:

- *ONE*: This level is the default setting in Cassandra. It returns a response from one replica for both read and write. For read, the replicas may not always have the most recent write, which means that you have to tolerate reading stale data. For write, the operation should be successfully performed on at least one replica. This strategy provides high level of availability and low level of consistency. Apparently, Cassandra prefers availability to consistency by default.
- *Write ALL*: This level writes to all replicas and read from one of them. A write will fail if a replica doesn't make a response. This level provides high level of consistency and read availability, but low level of write availability.
- *QUORUM*: For write, the operation must be successfully performed on more than half of replicas. For read, this level returns the record with the most recent timestamp after more than half of replicas have responded. This strategy provides high level of consistency and strong availability.

Similar to HBase, Cassandra updates are first written to a commit log stored on hard drive and then to an in-memory table called *memtable*. *Memtables* are periodically written to replicas stored on the disk. Cassandra determines the responsible replicas in fixed order when handling requests. The first replica, also called main replica, is always performed, no matter which consistency level is used.

---

[1] HBase also supports replicating data across datacenter for disaster recovery purpose only.

# 3 Benchmarking Replication and Consistency

We conduct several benchmarking efforts to give performance results for different replication/consistency strategies. These results are obtained through different workloads under specified system stress level. This work is based on YCSB [7]: Yahoo! Cloud Serving Benchmark. YCSB has become the de facto industry standard benchmark for cloud serving databases since its release. The core workloads in YCSB are sets of read/insert/update/scan operations mixed in different proportions. The operated records are selected use some distributions. Such workloads are apt to test the tradeoff strategies in different scenarios. What's more, YCSB is good at extensibility: Researchers can implement its interfaces to put new databases into this benchmark. To date, NoSQL databases like HBase, Cassandra, PNUTS [8], Voldemort [9] and many others have been benchmarked using YCSB.

## 3.1 Benchmark Methodology

In this work, all the tests are conducted on the same testbed to obtain comparative performance of databases using different replication and consistency strategies. In order to get credible and reasonable test results, several additional considerations should be carefully taken:

- *The number of test threads:* The number of client test threads of YCSB should be carefully chosen to prevent side effects of latencies caused by clients. If we use a heavy benchmark workload but a small number of test threads, each thread will be burdened with too many requests, as a result of which, the request latency rises for non-database-related reasons. The right number of client test threads can be set by analyzing the average system load of the client.
- *The number of records:* The number of records should be large enough to avoid a *local trap*. The local trap means that most of the operations are handled by only a few cluster nodes, as a result of which, we cannot obtain the overall performance of the cluster. This issue usually happens when there are not enough test data. The small number of records will also cause the *fit-in-memory* problem: The majority of the test data is cached in memory, as a result of which, benchmarking read performance become meaningless. The proper number of records should ensure that the test operations can be performed with disk access on the whole database cluster evenly.
- *The number of operations:* The number of operations should be large enough to generate substantial and stable load across all nodes evenly and make sure that the test can run for a long time to overcome side effects of *cold start* and *memory garbage collect*.

## 3.2 Benchmark Types

What make a benchmarking effort reasonable, reliable and meaningful? The reasonability and reliability of a benchmarking effort can be guaranteed by including some micro tests, because univariate results from such tests can be used to predict and explain results of comprehensive tests. The meaningfulness of a benchmarking effort can be archived by doing some stress tests, for stress tests can give an overview of the performance of the database system and are more similar to reality. Therefore, we introduce the following benchmarks into this effort:

- *Micro benchmark for replication:* This benchmark uses different replication strategies to compare the throughput and latency of databases. In order to get the most basic aspects of the performance, this benchmark uses workloads consisting of atomic operations. In order to reduce the latency variance introduced by the various size of transaction data, the test data consists of records of tiny size.
- *Stress benchmark for replication:* This benchmark uses different replication strategies to compares the throughput and latency of databases running at full speed. This benchmark can present an overview of the system performance and is more similar to reality for its scenario-based workloads.
- *Stress benchmark for consistency:* This benchmark uses different consistency levels to check out the changes in database throughputs. The workload of this benchmark is the same as that of the stress benchmark for replication.

### 3.3 Benchmark Workloads

As we mentioned above, the workloads in micro tests and stress tests are different. In micro tests, the test data consists of 1 billion records of 1 byte, and the workloads are basic insert/read/update/scan operations. In stress tests, the test data consists of 100 million records of 1000 bytes, the target records are chosen using some distributions, and the workloads are borrowed from YCSB with adjustments on the insert/read/update/scan ratio to simulate the real life workloads of different scenarios:

- *Read mostly:* This workload consists of reads mixed with a small portion of writes. This workload represents real life applications like reading on social website mixed with remarking actions.
- *Read latest:* This workload reads the newest updated records. The typical usage scenario is reading feeds on Twitter, Google plus etc.
- *Read & update:* This workload concerns reads and updates equally. The typical real life representative is the online shopping cart: People review the cart and change their choices.
- *Read-modify-write:* This workload reads some records to modify, then write them back. A typical real life usage is that people often review and change their profiles on websites.
- *Scan short range:* This workload retrieves records satisfying certain conditions. A typical application scenario is that people view news retrieved by recommended trends or topics on the social media websites.

We conclude the benchmark workloads used in the stress benchmark in Table 1.

### 3.4 Testbed

We use 16 server-class machines in the same rack as the testbed, which can reduce inferences from the partition problem (*a.k.a.* the P in the CAP theorem). Each machine

**Table 1.** Workloads of the stress benchmarks for replication and consistency

| Workload | Typical usage | Operations | Records distribution |
|---|---|---|---|
| Read mostly | Online tagging | Read/update ratio: 95/5 | *Zipfian* |
| Read latest | Feeds reading | Read/insert ratio: 80/20 | *Latest* |
| Read & update | Online shopping cart | Read/update ratio: 50/50 | *Zipfian* |
| Read-modify-write | User profile | Read/read-modify-write ratio: 50/50 | *Zipfian* |
| Scan short ranges | Topic retrieving | Scan/insert ratio: 95/5 | *Zipfian* |

owns two Xeon L5640 64 bit processers (each processer owns 6 cores and each core owns 2 threads), 32 GB of RAM, one hard drive and gigabit ethernet connection. In HBase tests, we configure 15 nodes as HRegion servers, leaving the last node serving as both HMaster and the YCSB client. In Cassandra tests, we also use 15 nodes to do the server job, leaving one machine to emit the test requests. We run Cassandra 2.0.8 and HBase 0.96.1.1 with recommended configurations in these tests. In order to make YCSB compatible with Cassandra 2.0.8 and HBase 0.96.1.1, we compile YCSB from the latest source code with modifications on library dependencies.

# 4    Experimental Results

The experimental results are shown and analyzed separately for different benchmarks.

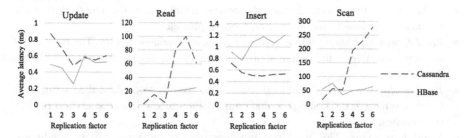

**Fig. 1.** Results of the micro benchmark for replication in HBase and Cassandra

## 4.1    Micro Benchmark for Replication

In this benchmark, we keep the load of the testbed in unsaturated state by limiting the number of concurrence requests, and conduct six rounds of testing. In each round, the replication factor is increased by one, and the update/read/insert/scan test is run one after another. Our expectations on changes in latency when the replication factor increases are:

- The read/scan latency changes slightly in both HBase and Cassandra tests. This is because no matter how many replicas exist, either HBase or Cassandra can only read from the main replica when using the default consistency strategy.
- The insert/update latency becomes higher in HBase tests. This is because HBase need to guarantee writing to all replicas successfully, and the write overhead become heavier when the replication factor increases.
- The insert/update latency changes slightly in Cassandra tests. This is because no matter how many replicas exist, Cassandra only need to guarantee writing to one replica successfully when using the default consistency strategy.

The experimental results are illustrated in Fig. 1, from which we can learn:

- Both the curve of read/scan latency in HBase tests and the curve of insert/update latency in Cassandra tests fluctuate smoothly, which is in line with the expectations.
- There is no significant change in the insert/update latency in HBase tests, which do not meet the expectation. The possible reason for such results is that when HBase writes, it do in-memory data replication instead of writing to replicas on hard drive instantly, which significantly reduce the write overhead.
- The read/scan latency in Cassandra tests increases rapidly as the replication factor is higher than 3, which is beyond the expectation. Such a result may be due to the extra burden introduced by the *read repair* process which issues writes to the out-of-date replicas in background to ensure that frequently-read data remains

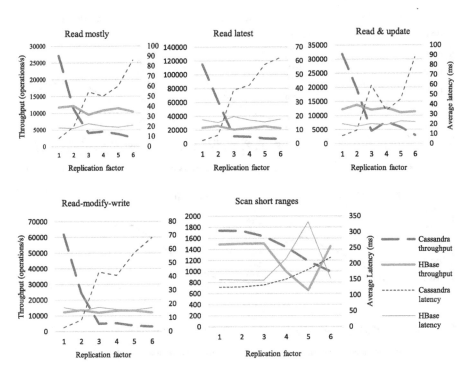

**Fig. 2.** Results of the stress benchmark for replication in HBase and Cassandra

consistent. Cassandra enables this feature by default [10], and the read-after-write test pipeline can trigger the *read repair* processes. When the replication factor increases, the burden of *read repair* continue to increase, which results in higher read latencies.

## 4.2   Stress Benchmark for Replication

In this benchmark, we use a constant number of test threads and a variety of target throughputs to detect the peak runtime throughput and the corresponding latency of databases. We conduct six rounds of testing. In each round, the replication factor is increased by one, and the read latest/scan short ranges/read mostly/read-modify-write/ read & update test is run one after another. Our expectations on changes in latency and throughput when the replication factor increases are:

- The runtime throughput is inverted-related with the latency in all tests. This is because when we use stress workloads to exam the upper limit of processing capacity, the latency depends on the capacity of the database cluster. The YCSB client will not emits a new request until it receives a response for the prior request — higher latencies will slow down the request emitting rate and then lead to lower runtime throughputs.

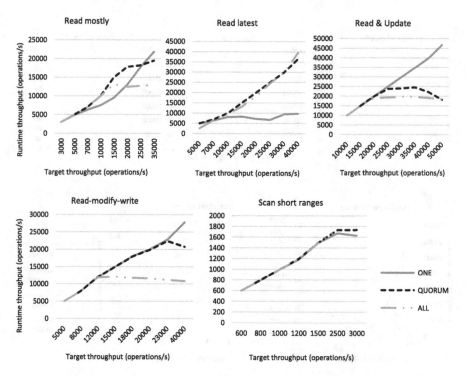

**Fig. 3.** Results of the stress benchmark for consistency in Cassandra

- The latency in all HBase tests changes insignificantly, which can be derived from the results of the HBase micro tests: All the five stress workloads simply consist of basic read/write operations. If the read/write latency in micro tests changes slightly, there is no reason for dramatic changes in latency in the stress tests.
- The latency in all Cassandra tests increases significantly, which can be derived from the changes of the read/scan latency in the Cassandra micro tests, because in all the stress workloads, at least 50 percent of operations are reads.

The throughput/latency versus replication factor results for HBase and Cassandra for the five stress workloads are illustrated in Fig. 2, from which we can learn that all the test results are in line with the expectations.

### 4.3   Stress Benchmark for Consistency

In this benchmark, we use a replication factor of 3, a constant number of test threads and a variety of target throughputs to detect the runtime throughput of Cassandra. There is no convenient method to adjust the default consistency strategy of HBase, hence we can only use Cassandra to do this job. Cassandra allows specifying the consistency level in request time, which makes the tests feasible. We conduct three rounds of testing, the consistency levels of which are respectively ONE, write ALL and QUORUM, and the read latest/scan short ranges/read mostly/read-modify-write/read & update test is run one after another in each round. Our expectations on changes in runtime throughput when the consistency level changes are:

- In the read latest test, level ONE performs worst, and level QUORUM and level ALL perform closely better. This is because the *read repair* process is frequently triggered in the read latest test. In this test, reads are intensively performed on new written records. Each write produces two inconsistent replicas in level ONE and almost zero inconsistent replica in level QUORUM/ALL. Apparently, more inconsistent replicas lead to heavier overhead of the *read repair* process.
- In the scan short ranges test, all the three levels perform closely. This is because overhead of the *read repair* process dramatically declined in this test, for we run this test after the read latest test which has repaired the majority of inconsistency. Moreover, there are only reads in this test, and in the perspective of YCSB client, reading from which of the replicas in the same rack is indifferently, as a result of which, the read latency can hardly be affected by the number of replicas too.
- In other tests, level ONE, level QUORUM and level ALL perform best, almost worst and worst respectively. This is because the write overhead becomes heavier when using a higher consistency level.

Figure 3 presents the runtime throughput versus target throughput with different consistency levels. We observe that all the expectations are confirmed by the experiment results with narrow biases. What's more, the bigger write proportion, the more obvious performance difference in these tests.

# 5  Related Work

It is complicated to conduct benchmarking efforts in the cloud serving database field: There are many tradeoffs need to think about when we evaluate a specific aspect of cloud serving databases. However, the YCSB framework brought a relatively easy way to do *apple-to-apple* comparisons between cloud serving databases, and has inspired some other benchmark tools in this field. The BigDataBench [11, 12], for example, has adopted YCSB as one of its components with extensions like the new metric on energy consumption.

With the help of YCSB, several meaningful efforts have been done in this filed. Pokluda et al. [13] benchmarked the availability of the failover characteristics of two representative systems: Cassandra and Voldemort. They used YCSB to monitor the throughput and latency of individual requests during a node failed and came back online, and found that transaction latency increased slightly while the node was down and that recovery. Bermbach et al. [14] evaluated the effects of geo-distributed replicas on the consistency of two cloud serving database: Cassandra and MongoDB. They used YCSB to generate workloads and Amazon EC2 to deploy these two databases. Replicas were distributed in same/different regions (Western Europe, northern California and Singapore) to compare the degree of inconsistency.

However, there are few efforts like this work to benchmark the replication factor.

# 6  Future Work

In this work, we specify the stress level using different target throughputs, which is inaccurate and lack of versatility for the comparison between different clusters. Another way to specify the stress level is using the service level agreement, SLA. An SLA is commonly specified like this: At least $p$ percentage of requests get response within $l$ latency during a period of time $t$. Using the SLA, We can keep user experiences at same level to compare throughputs of different systems. However, it is hard to specify an SLA using YCSB. We need to extend it.

There is *cold start* problem when we run benchmark workloads using YCSB, which leads to inaccurate results in latency tests. We have to run the tests for a long time, and repeat the tests several times to overcome this flaw. Consequently, the whole running time become long, and the benchmarking effort become inefficient in energy consumption. We need to optimize the method of test and result measurement used in YCSB.

Furthermore, this work has shown that a single rack of nodes cannot form a convincing testbed for more complicated tests such as geo-read latency test, partition test and availability test. We need to build a geo-distributed testbed to conduct such tests.

# 7  Conclusion

In this paper, we present our benchmarking effort on the replication and consistency strategies used in two cloud serving databases: HBase and Cassandra. This work is motivated by the tradeoff between latency, consistency and data replication. This benchmarking effort consists of three parts: Firstly, we use the atomic read/insert/update/scan operations to do micro tests to fetch the basic performance aspects of the target database. This part makes the foundation of further comprehensive benchmarks. Secondly, we change the replication factor to compare the performance of cloud serving databases. This part sketches an overview of the whole system performance. Finally, we use different consistency levels to compare the runtime throughputs. This part figures out the proper consistency strategy for some scenarios. We have observed some interesting benchmark results from a single-rack testbed:

- More replicas can hardly accelerate the read/write operation (in HBase), and even harm the read performance (In Cassandra using low level of consistency).
- The write latency dramatically increases when using higher level of consistency in Cassandra.
- High consistency level is not suitable for read latest and write heavy scenarios in Cassandra.

These results can give answers to the questions we mentioned in the introduction part within a SINGLE-RACK scope of validity.

In this effort, we also find the testbed is not suitable for read latency tests, and the benchmark tool has some efficiency affecting flaws. We will optimize the testbed and benchmark tools to conduct more rational and stable tests in the future.

# References

1. Chang, F., et al.: Bigtable: a distributed storage system for structured data. ACM Trans. Comput. Syst. (TOCS) **26**(2), 4 (2008)
2. Apache HBase. http://hbase.apache.org/
3. Apache Cassandra. http://cassandra.apache.org/
4. Brewer, E.: CAP twelve years later: How the "rules" have changed. Computer **45**(2), 23–29 (2012)
5. Abadi, D.J.: Consistency tradeoffs in modern distributed database system design. IEEE Comput. Mag. **45**(2), 37 (2012)
6. HBase read high-availability using timeline-consistent region replicas. http://issues.apache.org/jira/browse/HBASE-10070
7. Cooper, B.F., et al.: Benchmarking cloud serving systems with YCSB. In: Proceedings of the 1st ACM Symposium on Cloud Computing. ACM (2010)
8. Cooper, B.F., et al.: PNUTS: Yahoo!'s hosted data serving platform. Proc. VLDB Endow. **1**(2), 1277–1288 (2008)
9. Project Voldemort: A distributed database. http://www.project-voldemort.com/voldemort/
10. About Cassandra's Built-in Consistency Repair Features. http://www.datastax.com/docs/1.1/dml/data_consistency#builtin-consistency
11. Gao, W., et al.: Bigdatabench: a big data benchmark suite from web search engines (2013). arXiv preprint arXiv:1307.0320
12. Wang, L., et al.: Bigdatabench: A big data benchmark suite from internet services (2014). arXiv preprint arXiv:1401.1406

13. Pokluda, A., Sun, W.: Benchmarking Failover Characteristics of Large-Scale Data Storage Applications: Cassandra and Voldemort. http://www.alexanderpokluda.ca/coursework/cs848/CS848%20Project%20Report%20-%20Alexander%20Pokluda%20and%20Wei%20Sun.pdf

14. Bermbach, D., Zhao, L., Sakr, S.: Towards comprehensive measurement of consistency guarantees for cloud-hosted data storage services. In: Nambiar, R., Poess, M. (eds.) TPCTC 2013. LNCS, vol. 8391, pp. 32–47. Springer, Heidelberg (2014)

# Topical Section Headings:
# Workload Characterization

# I/O Characterization of Big Data Workloads in Data Centers

Fengfeng Pan[1,2(✉)], Yinliang Yue[1], Jin Xiong[1], and Daxiang Hao[1]

[1] Institute of Computing Technology, Chinese Academy of Sciences, Beijing, China
[2] University of Chinese Academy of Sciences, Beijing, China
panfengfeng@ncic.ac.cn

**Abstract.** As the amount of data explodes rapidly, more and more organizations tend to use data centers to make effective decisions and gain a competitive edge. Big data applications have gradually dominated the data centers workloads, and hence it has been increasingly important to understand their behaviour in order to further improve the performance of data centers. Due to the constantly increased gap between I/O devices and CPUs, I/O performance dominates the overall system performance, so characterizing I/O behaviour of big data workloads is important and imperative.

In this paper, we select four typical big data workloads in broader areas from the BigDataBench which is a big data benchmark suite from internet services. They are Aggregation, TeraSort, Kmeans and PageRank. We conduct detailed deep analysis of their I/O characteristics, including disk read/write bandwidth, I/O devices utilization, average waiting time of I/O requests, and average size of I/O requests, which act as a guide to design highperformance, low-power and cost-aware big data storage systems.

## 1 Introduction

In recent years, big data workloads [20] are more and more popular and play an important role in enterprises business. There are some popular and typical applications, such as TeraSort, SQL operations, PageRank and K-means. Specifically, TeraSort is widely used for page or document ranking; SQL operations, such as join, aggregation and select, are used for log analysis and information extraction; PageRank is widely used in search engine field; and Kmeans is usually used as electronic commerce algorithm.

These big data workloads run in data centers, and their performance is critical. The factors which affect their performance include: algorithms, hardware including node, interconnection and storage, and software such as programming model and file systems. This is the reason for why several efforts have been made to analyse the impact of these factors on the systems [17,20]. However, data access to persistent storage usually accounts for a large part of application time because of the ever-increasing performance gap between CPU and I/O devices.

© Springer International Publishing Switzerland 2014
J. Zhan et al. (Eds.): BPOE 2014, LNCS 8807, pp. 85–97, 2014.
DOI: 10.1007/978-3-319-13021-7_7

With the rapid growth of data volume in many enterprises and a strong desire for processing and storing data efficiently, a new generation of big data storage system is urgently required. In order to achieve this goal, a deep understanding of big data workloads present in data centers is necessary to guide the big data systems design and tuning.

Some studies have been conducted to explore the computing characteristics of big data workloads [16,18], meanwhile, lots of work also have been done to depict the storage I/O characteristics of enterprise storages [5,9]. However, to the best of our knowledge, none of the existing research has understood the I/O characteristics of big data workloads, which is much more important in the current big data era. So understanding the characteristics of these applications is the key to better design the storage system and optimize their performance and energy efficiency.

In this paper, we choose four typical big data workloads as mentioned above, because they have been widely used in popular application domains [1,20], such as search engine, social networks and electronic commerce. Detailed information about I/O metrics and workloads are shown in Sect. 3.2. Through the detailed analysis, we get the following four observations. First, the change of the number of task slots has no effects on the four I/O metrics, but increasing the number of task slots appropriately can accelerate the process of application execution. Second, increasing memory can alleviate the pressure of disk read/write, and effectively improve the I/O performance when the data size is large. Third, the compression of intermediate data mainly affects the MapReduce I/O performance and has little influence on HDFS I/O performance. However, compression consumes some CPU resource which may influence the job's execution time. Fourth, the I/O pattern of HDFS and MapReduce are different, namely, HDFS's I/O pattern is large sequential access and MapReduce's I/O pattern is small random access, so when configuring storage systems, we should take several factors into account, such as the number of devices and the types of devices.

The rest of the paper is organized as follows: Sect. 2 discusses related work. Section 3 describes the experimental methodology. Section 4 shows the experimental results. Section 5 briefly brings up future work and concludes this paper.

## 2    Related Work

Workloads characterization studies play a significant role in detecting problems and performance bottlenecks of systems. Many workloads have been extensively studied in the past, including enterprise storage systems [7,10], web server [15,19], HPC cluster [14] and network systems [12].

### 2.1    I/O Characteristics of Storage Workloads

There have been a number of papers about the I/O characteristics of storage workloads [8,11,15].

Kavalanekar et al. [15] characterized large online services for storage system configuration and performance modeling. It contains a set of characteristics, including block-level statistics, multi-parameter distributions and rankings of file access frequencies. Similarly, Delemitrou et al. [11] presented a concise statistical model which accurately captures the I/O access pattern of large-scale applications including their spatial locality, inter-arrival times and type of accesses.

### 2.2 Characteristics of Big Data Workloads

Big data workloads have been studied in recent years at various levels, such as job characterization [16–18], storage systems [5,9].

Kavulya et al. [16] characterized resource utilization patterns, job patterns, and source of failure. This work focused on predicting job completion time and found the performance problems. Similarly, Ren et al. [18] focused on not only job characterization and task characterization, but also resource utilization on a Hadoop cluster, including CPU, Disk and Network.

However, the above researches on big data workloads focused on job level, but not on the storage level. Some studies have provided us with some metrics about data access pattern in MapReduce scenarios [4,6,13], bur these metrics are limited, such as block age at time of access [13] and file popularity [4,6]. Abad et al. [5] conducted a detailed analysis about some HDFS characterization, such as file popularity, temporal locality, request arrival patterns, and then figure out the data access pattern of two types of big data workloads, namely batch and interactive query workloads. But this work concentrates on HDFS's I/O characterization, does not study the intermediate data, and also this work only involves two types of workloads.

In this paper, we focus on I/O characteristics of big data workloads, the difference between our work and the previous work is that first, we focus on the I/O behaviour of individual workload and choose four typical workloads in broader areas, including search engine, social network, e-commerce, which are popular in the current big data era; Second, we analyze the I/O behavior of both HDFS and MapReduce intermediate data from different I/O characteristics, including disk read/write bandwidth, I/O devices' utilization, average waiting time of I/O requests, and average size of I/O requests.

## 3  Experimental Methodology

This section firstly describes our experiment platform, and then presents workloads used in this paper.

### 3.1  Platform

We use an 11-node (one master and ten slaves) Hadoop cluster to run the four typical big data workloads. The nodes in our Hadoop cluster are connected through 1 Gb ethernet network. Each node has two Intel Xeon E5645 (Westmere)

**Table 1.** The detailed hardware configuration information

| CPU Type | Intel R Xeon E5645 |
|---|---|
| # Cores | 6 cores@2.4 G |
| # threads | 12 threads |
| Memory | 32 GB, DDR3 |
| Disk | **6 disks** (one disk for system, three disks for HDFS data, the other two disks for MapReduce intermediate data) |
| | **Disk Model Seagate**: ST1000NM0011 |
| | **Capacity**: 1TB |
| | **Rotational Speed**: 7200 RPM |
| | **Avg. Seek/Rotational Time**: 8.5 ms/4.2 ms |
| | **Sustained Transfer Rate**: 150 MB/s |

**Table 2.** The detailed software configuration information

| Hadoop | 1.0.4 |
|---|---|
| JDK | 1.6.0 |
| Hive | 0.11.0 |
| OS Distribution and Linux kernel | Centos 5.5 with the 2.6.18 Linux kernel |
| TeraSort | BigDataBench2.1 [2] |
| SQL operations | BigDataBench2.1 |
| PageRank | BigDataBench2.1 |
| Kmeans | BigDataBench2.1 |

processors, 32 GB memory and 7 disks(1TB). A Xeon E5645 processor includes six physical out-of-order cores with speculative pipelines. Tables 1 and 2 shows the detailed configuration information.

### 3.2 Workloads and Statistics

We choose four popular and typical big data workloads from BigDataBench [20]. BigDataBench is a big data benchmark suite from internet services and it provides several big data generation tools to generate various types and volumes of big data from small-scale real-world data while preserving their characteristics. Table 3 shows the description of the workloads, which is characterized in this paper.

Iostat [3] is a well-used monitor tool used to collect and show various system statistics, such as CPU times, memory usage, as well as disk I/O statistics. In this paper, we mainly focus on the disk-level I/O behaviour of the workloads, and we extract information from iostat's report, and the metrics which we focus on are shown in Table 4.

**Table 3.** The description of the workloads

| Workloads | Performance bottleneck | Scenarios | Input Data size |
|-----------|------------------------|-----------|-----------------|
| TeraSort (TS) | I/O bound | Page ranking; document ranking | 1 TB |
| Aggregation (AGG) | CPU bound | Log analysis; information extraction | 1 TB |
| K-means (KM) | CPU bound in iteration; I/O bound in clustering | Clustering and Classification | 512 GB |
| PageRank (PR) | CPU-bound | Search engine | 512 GB |

**Table 4.** Notation of I/O characterization

| I/O characterization | Description | Notes |
|----------------------|-------------|-------|
| rMB/s and wMB/s | The number of megabytes read from or written to the device per second | Disk Read or Write Bandwidth |
| %util | Percentage of CPU time during which I/O requests were issued to the device | Disk utilization |
| await (ms) | The average time for I/O requests issued to the device to be served. This includes the time spent by the requests in queue and the time spent servicing them | average waiting time of I/O request = await - svctm |
| svctm (ms) | The average service time for I/O requests that were issued to the device | |
| avgrq-sz (the number of sectors) | The average size of the requests that were issued to the device. And the size of sectors is 512B | average size of I/O request |

# 4 Results

In this section, we describe HDFS/MapReduce I/O characteristics from four metrics, namely, disk read/write bandwidth, I/O devices' utilization, average waiting time of the I/O requests, and average size of the I/O requests. Through the comparison of each workloads, we can obtain the I/O characterization of HDFS and MapReduce respectively, and also the difference between HDFS and MapReudce.

In addition, different Hadoop configurations can influence the workloads' execution. So in this paper, we select three factors and analyse their impact on the I/O characterization of these workloads. The three factors are as follows. First, *the number of task slots, including map slots and reduce slots.* In Hadoop, computing resource is represented by slot, there are two types of slot: map task slot and reduce task slot. Here computing resource refers to CPU.

Second, *the amount of physical memory of each node.* As we know that memory plays an important role in I/O performance, so memory is an important factor which affects the I/O behaviour of workloads. So, the second factor we focus on is the relationship be-tween memory and I/O characterization. Third, *whether the intermediate data is compressed or not.* Compression involves two types of resources: CPU and I/O. what's the influence on the I/O behaviour of workloads when CPU and I/O resources both change. So, the final factor we focus on is the relationship between compression and I/O characterization.

## 4.1    Disk Read/Write Bandwidth

**Task Slots.** Figure 1 shows the effects of the number of task slots on the disk read/write bandwidth in HDFS and MapReduce respectively. In these experiments, each node configured 16 GB memory and the intermediate data is compressed. "10_8", "20_16" in the figures mean the number of map task slots and reduce task slots respectively.

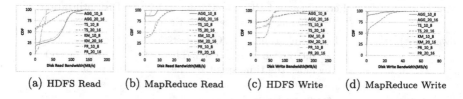

(a) HDFS Read     (b) MapReduce Read     (c) HDFS Write     (d) MapReduce Write

**Fig. 1.** The effects of the number of task slots on the Disk Read/Write Bandwidth in HDFS and MapReduce respectively

From Fig. 1 we can get the following two conclusions. First, when the number of task slots changes, there is barely any influence on the disk read/write bandwidth in both scenarios for every workload. Second, the variation of disk read/write bandwidth of different workloads in both scenarios are disparate because the data volume of each workload in different phases of execution are not the same.

In a word, there is little effect on disk read/write bandwidth when the number of task slots changes. However, configuring the number of task slots appropriately can reduce the execution time of workloads, so we should take it into account when workloads run.

**Memory.** Figure 2 displays the effects of the memory size on the disk read/write bandwidth in HDFS and MapReduce respectively. In these experiments, task slots configuration on each node is 10_8 and the intermediate data is not compressed. "16G", "32G" in the figures mean the memory size of node.

As Fig. 2 shows, the influence of the memory size on disk read/write Bandwidth depends on the data volume. There is a growth of disk read bandwidth

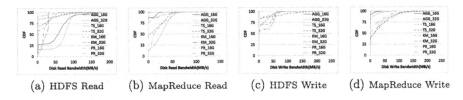

(a) HDFS Read      (b) MapReduce Read      (c) HDFS Write      (d) MapReduce Write

**Fig. 2.** The effects of memory on the Disk Read/Write Bandwidth in HDFS and MapReduce respectively

in HDFS when memory increases as shown in Fig. 2(a) due to the large amount of raw data, but after handling and processing the raw data, the disk write band-width of each workloads in HDFS are different because of the final data volume. When the final data volume is small, memory has no effects on the disk write bandwidth, such as Kmeans, as shown in Fig. 2(c). This result can also be reflected in MapReduce.

**Compression.** Figure 3 exhibits the effects of intermediate data compression on the disk read/write bandwidth in MapReduce. In these experiments, each node configured 32 GB memory and task slots configuration is 10_8. "off", "on" in the figures mean whether the inter-mediate data is compressed or not.

As Fig. 3 shows, due to the reduction of intermediate data volume with compression, the disk read/write bandwidth increase in MapReduce.

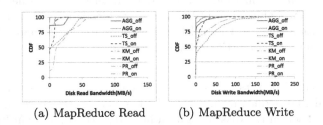

(a) MapReduce Read      (b) MapReduce Write

**Fig. 3.** The effects of compression on the Disk Read/Write Bandwidth in MapReduce.

In addition, compression has little impact on the HDFS's disk read/write bandwidth, so we do not present the result.

### 4.2   Disk Utilization

**Task Slots.** Figure 4 depicts the effects of the number of task slots on the disk utilization in HDFS and MapReduce respectively.

From Fig. 4 we can get the following two conclusions. First, the trends of workloads in disk utilization are the same when the number of task slots changes,

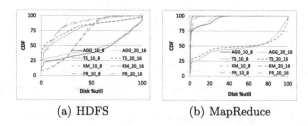

(a) HDFS          (b) MapReduce

**Fig. 4.** The effects of the number of task slots on the Disk Utilization in HDFS and MapReduce respectively.

**Table 5.** The HDFS/MapReduce Disk %util ratio

|     | >90 %uitl | >95 %uitl | >99 %util |
|-----|-----------|-----------|-----------|
| AGG | 22.6 % / 0 | 16.4 % / 0 | 9.8 % / 0 |
| TS  | 5.2 % / 27.2 % | 3.8 % / 15.6 % | 2.4 % / 5.5 % |
| KM  | 0.4 % / 0 | 0.3 % / 0 | 0.2 % / 0 |
| PR  | 0.5 % / 0.1 % | 0.3 % / 0.1 % | 0.2 % / 0.1 % |

i.e. the number of task slots has little impact on the disk utilization in both scenarios. Second, workloads have different behaviour about disk utilization in both scenarios. From the Table 5, the HDFS disk utilization of Aggregation is higher than the others, so Aggregation HDFS disk may be the bottleneck. Similarly, the MapReduce disk utilization of TeraSort is higher than the others; From the Fig. 4(b), the MapReduce disk utilization of workloads is 50 % or less at their most of execution time, so the disks are not busy, except TeraSort because of the large amount of TeraSort's intermediate data.

**Memory.** Figure 5 depicts the effects of the memory size on the disk utilization in HDFS and MapReduce respectively. As Fig. 5(a) shows, increasing memory size has no impact on the disk utilization in HDFS. However, from Fig. 5(b) we can see that there is a difference between HDFS and MapReduce. In MapReduce, the disk utilization of Aggregation and Kmeans has no changes when memory size changes because the devices are not busy before memory changes. However, the disk utilization of TeraSort and PageRank is reduced when memory increases as shown in Fig. 5(b). So, increasing memory size can help reduce the number of I/O requests and ease the bottleneck of disk.

**Compression.** From Fig. 6(a), we can know that when intermediate data is compressed, the HDFS's disk utilization essentially unchanges. As Fig. 6(b) shows, there is no influence on intermediate data's disk utilization of TeraSort, Aggregation and Kmeans, except PageRank in MapReduce.

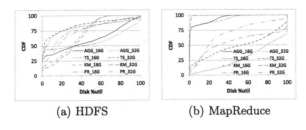

(a) HDFS                    (b) MapReduce

**Fig. 5.** The effects of memory on the Disk Utilization in HDFS and MapReduce respectively.

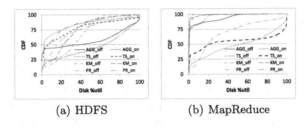

(a) HDFS                    (b) MapReduce

**Fig. 6.** The effects of compression on the Disk Utilization in HDFS and MapReduce respectively.

## 4.3   Average Waiting Time of I/O Request

**Memory.** Figure 7 shows the effects of Memory on the disk waiting time of I/O requests in HDFS and MapReduce respectively.

From Fig. 7, we can learn that the disk average waiting time of I/O requests of workloads varies with different memory size, in other words, the memory size has an impact on the disk waiting time of I/O requests and the MapReduce disk waiting time of I/O request is larger than the HDFS's.

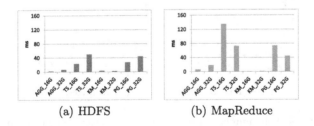

(a) HDFS                    (b) MapReduce

**Fig. 7.** The effects of memory on the Disk waiting time of I/O requests in HDFS and MapReduce respectively.

**Compression.** Figure 8 depicts the effects of compression on the disk waiting time of I/O requests in HDFS and MapReduce respectively.

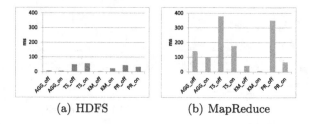

(a) HDFS                    (b) MapReduce

**Fig. 8.** The effects of compression on the Disk waiting time of I/O requests in HDFS and MapReduce respectively.

From Fig. 8, we can see that the disk average waiting time of I/O requests remains unchanged in HDFS because HDFS's data is not compressed, however, due to the reduction of intermediate data volume with compression, the disk waiting time of I/O request in MapReduce is decreased. And the MapReduce disk waiting time is larger than the HDFS's because of their different I/O mode in access pattern, i.e. HDFS's access pattern is domated by large sequential accesses, while MapReduce's access pattern is dominated by smaller random access. This result can be seen in Fig. 9.

### 4.4   Average Size of I/O Requests

**Task Slots.** Figure 9 depicts the effects of the number of task slots on the disk average size of I/O request in HDFS and MapReduce respectively.

As the task slots is a kind of computing resource, there is little impact on the disk average size of I/O requests when the number of task slots changes from the figures. Also, the average size of HDFS I/O requests is larger than the MapReduce's because they have different I/O mode in I/O granularity. In other words, HDFS's I/O granularity is larger than the MapReduce's.

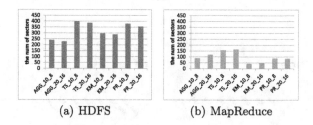

(a) HDFS                    (b) MapReduce

**Fig. 9.** The effects of the number of task slots on the Disk average size of I/O request in HDFS and MapReduce respectively.

The same result also can be achieved by the effects of memory on the disk average size of I/O requests. However, due to the compression, the effects of memory on the disk average size of I/O requests is different from the effects of the number of task slots on the disk average size which is reflected in Fig. 10.

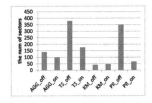

**Fig. 10.** The effects of compression on the Disk average size of I/O request in MapReduce.

**Compression.** In addition, whether the intermediate data is compressed or not has no impact on the HDFS's disk average size of I/O requests, so we do not present the result.

Figure 10 displays the effects of compression on the disk average size of I/O requests in MapReduce. It is seen from figure that as the intermediate data is compressed, the disk average size of I/O requests is decreased, and the percentage of reduction varies with the types of workloads due to the intermediate data volume. As Fig. 10 shows, there is little influence on the disk average size of I/O request when the intermediate data volume is small, such as Aggregation and Kmeans.

## 5  Conclusion and Future Work

In this paper, we have presented a study of I/O characterization of big data workloads. These workloads are typical, which are representative and common in search engine, social networks and electronic commerce. In contrast with previous work, we take into account disk read/write bandwidth, average waiting time of I/O requests, average size of I/O requests and storage device utilization, which are important for big data workloads. Some observations and implications are concluded as follows. First, task slots has little effects on the four I/O metrics, but increasing the number of task slots can accelerate the process of application execution. Second, the compression of intermediate data mainly affects the MapReduce I/O performance and has little influence on HDFS I/O performance. However, compression consumes some CPU resource which may influence the job's execution time. Third, increasing memory can alleviate the pressure of disk read/write and effectively improve the I/O performance when the data size is large. Last, HDFS data and MapReduce intermediate data have different I/O mode, which leads us to configuring their own storage systems according to their I/O mode.

**Acknowledgement.** This paper is supported by National Science Foundation of China under grants no. 61379042, 61303056, and 61202063, and Huawei Research Program YB2013090048.

# References

1. http://www.alexa.com/topsites/global
2. http://prof.ict.ac.cn/BigDataBench/
3. http://linux.die.net/man/1/iostat
4. Abad, C.L., Lu, Y., Campbell, R.H.: Dare: adaptive data replication for efficient cluster scheduling. In: 2011 IEEE International Conference on Cluster Computing (CLUSTER), pp. 159–168 (2011)
5. Abad, C.L., Roberts, N.: A storage-centric analysis of mapreduce workloads: file popularity, temporal locality and arrival patterns. In: 2012 IEEE International Symposium on Workload Characterization (IISWC), pp. 100–109 (2012)
6. Ananthanarayanan, G., Agarwal, S.: Scarlett: coping with skewed content popularity in mapreduce clusters. In: Proceedings of the Sixth Conference on Computer Systems (2011)
7. Bairavasundaram, L.N., Arpaci-Dusseau, A.C., Arpaci-Dusseau, R.H., Goodson, G.R., Schroeder, B.: An analysis of data corruption in the storage stack. ACM Transactions on Storage (TOS) **4** (2008)
8. Kozyrakis, C., Kansal, A., Sankar, S., Vaid, K.: Server engineering insights for large-scale online services. IEEE Micro **30**, 8–19 (2010)
9. Chen, Y., Alspaugh, S., Katz, R.: Interactive analytical processing in big data systems: a cross-industry study of mapreduce workloads. In: Proceedings of the VLDB Endowment (2012)
10. Chen, Y., Srinivasan, K., Goodson, G.: Design implications for enterprise storage systems via multi-dimensional trace analysis
11. Delimitrou, C., Sankar, S., Vaid, K., Kozyrakis, C.: Decoupling datacenter studies from access to large-scale applications: a modeling approach for storage workloads. In: 2011 IEEE International Symposium on Workload Characterization (IISWC), pp. 51–60 (2011)
12. Ersoz, D., Yousif, M.S., Das, C.R.: Characterizing network traffic in a cluster-based, multi-tier data center. In: 27th International Conference on Distributed Computing Systems, ICDCS '07, p. 59 (2007)
13. Fan, B., Tantisiriroj, W., Xiao, L., Gibson, G.: Diskreduce: raid for data-intensive scalable computing. In: Proceedings of the 4th Annual Workshop on Petascale Data Storage (2009)
14. Iamnitchi, A., Doraimani, S., Garzoglio, G.: Workload characterization in a high-energy data grid and impact on resource management. In: 2009 IEEE International Conference on Cluster Computing (CLUSTER), pp. 100–109 (2009)
15. Kavalanekar, S., Worthington, B.: Characterization of storage workload traces from production windows servers. In: 2008 IEEE International Symposium on Workload Characterization (IISWC), pp. 119–128 (2008)
16. Kavulya, S., Tan, J., Gandhi, R., Narasimhan, P.: An analysis of traces from a production mapreduce cluster. In: 2010 10th IEEE/ACM International Conference on Cluster, Cloud and Grid Computing (CCGrid), pp. 94–103 (2010)
17. Kyrola, A., Blelloch, G., Guestrin, C.: Graphchi: large-scale graph computation on just a pc. In: Proceedings of the 10th USENIX Conference on Operating Systems Design and Implementation (2012)
18. Ren, Z., Xu, X., Wan, J., Shi, W., Zhou, M.: Workload characterization on a production hadoop cluster: a case study on taobao. In: 2012 IEEE International Symposium on Workload Characterization (IISWC), pp. 3–13 (2012)

19. Sankar, S., Vaid, K.: Storage characterization for unstructured data in online services applications. In: 2009 IEEE International Symposium on Workload Characterization (IISWC), pp. 148–157 (2009)
20. Wang, L., Zhan, J., Luo, C., et al.: Bigdatabench: a big data benchmark suite from internet services. In: 2014 IEEE 20th International Symposium on High Performance Computer Architecture (HPCA), pp. 488–499 (2014)

# Characterizing Workload of Web Applications on Virtualized Servers

Xiajun Wang[1,2], Song Huang[2], Song Fu[2(✉)], and Krishna Kavi[2]

[1] Department of Information Engineering,
Changzhou Institute of Light Industry Technology, Changzhou, China
dy_wxj@163.com
[2] Department of Computer Science and Engineering,
University of North Texas, Denton, TX, USA
SongHuang@my.unt.edu, {Song.Fu,Krishna.Kavi}@unt.edu

**Abstract.** With the ever increasing demands of cloud computing services, planning and management of cloud resources has become a more and more important issue which directed affects the resource utilization and SLA and customer satisfaction. But before any management strategy is made, a good understanding of applications' workload in virtualized environment is the basic fact and principle to the resource management methods. Unfortunately, little work has been focused on this area. Lack of raw data could be one reason; another reason is that people still use the traditional models or methods shared under non-virtualized environment. The study of applications' workload in virtualized environment should take on some of its peculiar features comparing to the non-virtualized environment. In this paper, we are open to analyze the workload demands that reflect applications' behavior and the impact of virtualization. The results are obtained from an experimental cloud testbed running web applications, specifically the RUBiS benchmark application. We profile the workload dynamics on both virtualized and non-virtualized environments and compare the findings. The experimental results are valuable for us to estimate the performance of applications on computer architectures, to predict SLA compliance or violation based on the projected application workload and to guide the decision making to support applications with the right hardware.

**Keywords:** Workload characterization · Virtualization · Performance modeling · Cloud computing

## 1 Introduction

The increasingly popular cloud computing paradigm provides on-demand access to computing and storage with the appearance of unlimited resources [1]. Users are given access to a variety of data and software utilities to manage their work. Users rent virtual resources and pay for only what they use. Underlying these services are data centers that provide virtual machines (VMs) [2]. Virtual machines make it easy to host computation and applications for large numbers of distributed users by giving each the illusion of a dedicated computer system. It is anticipated that cloud platforms and services will increasingly play a critical role in academic, government and industry sectors, and will have widespread societal impact.

© Springer International Publishing Switzerland 2014
J. Zhan et al. (Eds.): BPOE 2014, LNCS 8807, pp. 98–108, 2014.
DOI: 10.1007/978-3-319-13021-7_8

Resource planning and management is crucial for building cost-effective cloud systems and services with a high service-level agreement (SLA) and customer satisfaction rate. Current solutions to resource management usually over-provision VMs and/or their capacity to cloud applications [3]. However, a fundamental question, i.e., "What are the characteristics of applications' runtime behavior on the cloud?" or "What impact does virtualization have on the resource demands from cloud applications?", has not yet been answered. There exists research on analyzing the performance traces collected from data centers [4, 5]. Still, none of them evaluate the influence of virtualization on the applications' resource demands in cloud computing infrastructures.

The goal of this work is to characterize runtime workload of cloud applications in the virtualized environment and compare it with traditional, non-virtualized systems. To the best of our knowledge, this is the first work to analyze the impact of virtualization on the resource demands of cloud applications. In this paper, we present the experimental results on a cloud testbed. We run an illustrating web application, i.e., RUBiS (Rice University Bidding System) benchmark [6], on cloud servers. We profile the application's workload dynamics on both virtualized and non-virtualized environments. We compare the resource demands of CPU, RAM, disk and network at the three tiers (i.e., web, application and database servers) of RUBiS while serving thousands of client requests. The findings and knowledge will help us accurately estimate the performance of applications, predict SLA compliance or violation based on the projected application workload and guide the decision making to support applications with the right hardware in the cloud.

The rest of this paper is organized as follows. Section 2 discusses the related work. We describe the settings of the cloud testbed and the application benchmark in Sect. 3. The experimental results are presented in Sect. 4. Section 5 concludes the paper with remarks on the future work.

## 2 Related Work

Workload characterization studies are useful for helping system operators identify system bottlenecks and design solutions for performance optimization. Existing research efforts target different systems and components including data centers [4, 5], Web servers [7, 8], storage [9–11] and network [12, 13]. Several studies [14–16] focus on workload analysis in the grid and parallel computing systems. They present various methods for analyzing and modeling workload traces. However, the application characteristics and resource scheduling policies in high-performance computing (HPC) systems are different from those in the cloud [17–19].

Existing work on workload characterization can be classified into two major categories: model-driven and trace-driven methods. Model-driven approaches, such as [20], analyze resource utilization and application performance based on assumptions of workload distributions. The resource demand of a program is estimated by checking the types and number of instructions of the program and its structure. The overhead of modeling large and complex applications is prohibitive and the accuracy of the models is compromised by static analysis.

Trace-driven approaches study performance traces collected from real or controlled systems in order to discover the time series of user requests and resource usage. Distributions of profiled metrics are analyzed to describe workload characterization. For example, Kavulya et al. [21] analyze the job patterns and failure sources based on application execution traces from an HPC cluster. Mishra et al. [22] focus on the characteristics of resource demands on CPU and memory. The Yahoo Cloud Serving Benchmark [23] characterizes the activity of database-like systems at the read/write level. Their work focus on estimating application completion time and looking for performance problems based on application execution traces. Moreover, as applications display various workload dynamics, it is difficult to exploit this approach in capacity planning and real system analysis.

There is little work on understanding applications' workload dynamics in cloud computing environments. As virtualization has been an enabling technology for cloud computing, it is imperative to investigate the impact of virtualization on the resource demands of cloud applications, which is the focus of this work.

## 3  Cloud Testbed and Benchmark

The cloud computing system under test consists of HP ProLiant servers which are connected by gigabit Ethernet. Each cloud server is equipped with 8 Intel Xeon 2.8 GHz cores, 32 GB of RAM and 2 TB of disk. We have installed Xen 3.1.2 hypervisors on the cloud servers. The operating system on a virtual machine is Linux 2.6.18 as distributed with Xen 3.1.2. The cloud testbed is organized and built in an Amzon EC2-like [24] style providing IaaS cloud services. Each cloud server hosts up to ten VMs. A VM is assigned up to two VCPUs, among which the number of active ones depends on applications. The amount of memory allocated to a VM is set to 2 GB.

On the cloud testbed, we run the RUBiS [6] distributed online web service benchmark as an illustrating cloud service. RUBiS provides an auction site prototype modeling eBay.com and it is widely used as the benchmark program to evaluate the server performance and web application designs. The RUBiS servers form a three-tier server architecture consisted of the Web, application and database servers. RUBiS clients send requests with different workload patterns (browsing, bidding and mixed with adjustable composition of the two actions) to the Web server and simulate auctions of items on eBay.

To profile the application's resource demands in the cloud environment, we exploit third-party monitoring tools, sysstat [25] to collect runtime performance data in the hypervisor and VMs, and a modified perf [26] to obtain the values of performance counters from the Xen hypervisor on each server in the cloud testbed. In total, 518 metrics are profiled, i.e., 182 for the hypervisor and 182 for VMs by sysstat and 154 for performance counters by perf, periodically. They cover the statistics of every component of cloud servers, including the CPU usage, process creation, task switching activity, memory and swap space utilization, paging, interrupts, network activity, I/O and data transfer, power management, and more. Table 1 lists and describes a sampling of the performance metrics that are used to characterize the workload dynamics of cloud applications on our testbed.

**Table 1.** A sample of performance metrics used to characterize workload of the RUBiS benchmark system on the cloud testbed.

| Metric | Description |
|---|---|
| %system | Percentage of CPU utilization that occurred while executing at the system level. |
| %user | Percentage of CPU utilization that occurred while executing at the user level. |
| %nice | Percentage of CPU utilization that occurred while executing at the user level with nice priority. |
| %iowait | Percentage of time that the CPU or CPUs were idle during which the system had an outstanding disk I/O request. |
| %soft | Percentage of time spent by the CPU or CPUs to service software interrupts. |
| %steal | Percentage of time spent in involuntary wait by the virtual CPU or CPUs while the hypervisor was serving another virtual processor. |
| proc/s | Total number of tasks created per second. |
| cswch/s | Total number of context switches per second. |
| intr/s | Total number of interrupts received per second by the CPU. |
| kbmemused | Amount of used memory in kilobytes. |
| kbbuffer | Amount of memory used as buffers by the kernel in kilobytes. |
| %memused | Percentage of used memory. |
| pswpin/s | Total number of swap pages the system brought in per second. |
| pswpout/s | Total number of swap pages the system brought out per second. |
| pgpgin/s | Total number of kilobytes the system paged in from disk per second. |
| pgpgout/s | Total number of kilobytes the system paged out to disk per second. |
| fault/s | Number of page faults (major+minor) made by the system per second. |
| pgsteal/s | Number of pages the system has reclaimed from cache (pagecache and swapcache) per second to satisfy its memory demands. |
| pgscank/s | Number of page scanned by the kswapd daemon per second. |
| %vmeff | Calculated as pgsteal/pgscan, a metric of the efficiency of page reclaim. |
| await | The average time (in milliseconds) for I/O requests issued to the device to be served. |
| tps | Total number of transfer per second that were issued to physical devices. |
| wr_sec/s | Number of sectors written to the device. |
| rd_sec/s | Number of sectors read from the device. |
| rxpck/s | Total number of packets received per second. |
| txpck/s | Total number of packets transmitted per second. |
| tcp-tw | Number of TCP sockets in $TIME_{WAIT}$ state. |
| cycles | Total number of CPU cycles. |
| ITLB-load | Total number of load operations to the instruction TLB. |
| branches | Number of branch operations. |
| branch-misses | Percentage of branch misses with the total number of branches. |
| LLC-store-misses | Number of last-level cache store misses operations. |
| LLC-prefetches | Number of last-level cache prefetches operations. |
| Major-faults | Number of page major faults. |
| DTLB-load-misses | Number of load misses operation to the data TLB. |
| DTLB-stores | Number of stores operation to the data TLB. |

# 4   Experimental Results and Analysis

We run the RUBiS benchmark system on the cloud testbed and profile the workload dynamics with different clients' request patterns on both virtualized and non-virtualized environments. In this section, we present the results from the experiments and analyze them to find the workload characteristics and the impact of virtualization on the dynamics of resource demands.

## 4.1   Workload Characterization in a Virtualized Environment

In the first set of experiments, we deployed the RUBiS servers in VMs: the front-end Apache web server and PHP application server (The two servers are integrated together in the PHP implementation.) and the back-end MySQL database server. 1000 clients external to the cloud testbed sent browsing, bidding and mixed type requests to the web server. The think time was set to 7 s. We ran the experiments for around 20 min and profiled the resource demands for CPU, RAM, disk, network, TCP socket and context switch both in VMs and the hypervisor (*dom0*). Figures 1, 2, 3, 4, 5 and 6 depict the workloads. We tested five types of request compositions: browsing only, bidding only, 30 % browsing and 70 % bidding, 50 % browsing and 50 % bidding, and 70 % browsing and 30 % bidding. Due to the space limitation, we only include the results of the first two compositions in this paper.

The first two sub-figures in each set show the workload demands of the web and application servers and the database server for virtualized resources, including CPU cycles, amount of RAM, disk reads and writes, data received and transmitted through networks, number of TCP sockets and number of context switches in VMs. The last sub-figure in each set presents the overall workload demands to the physical resources.

From the figures, we can see the workload curves for different types of resources display different shapes/distributions with different means and variances. But for each type of resource, the workload dynamics show some patterns that can be quantified by formal models. In addition, there exist some lags between workload changes of the database server and the web and application servers as the client requests are received and processed first by the web server before being sent to the back-end database server.

Between the front-end servers and back-end server, the front-end servers generate higher workload demands as they demand 6.11, 3.29, 5.71, 55.56 and 1.85 times more CPU cycles, RAM space, disk read/write, network data and context switches than the back-end server, but the number of TCP sockets is almost same. When we compare the aggregated workload demands of the VMs with that of the hypervisor, the former is 16.84, 0.58, 0.47, 0.98, 2.67 and 0.09 times more/less than the latter with regard to the six types of resources. This indicates the hypervisor performs additional work other than the workload of RUBiS servers.

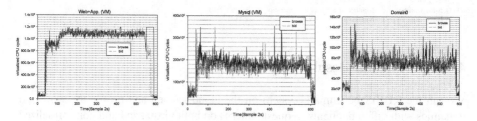

**Fig. 1.** CPU cycle demands by the web and application servers and the database servers in VMs and the hypervisor (*dom0*) to process the browsing and bidding requests from 1000 clients.

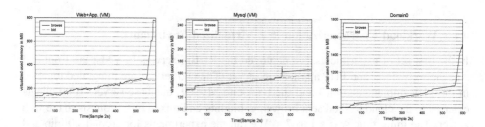

**Fig. 2.** RAM demands by the web and application servers and the database servers in VMs and the hypervisor.

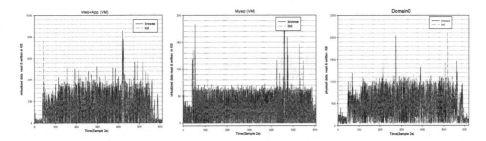

**Fig. 3.** Disk read and write by the web and application servers and the database servers in VMs and the hypervisor.

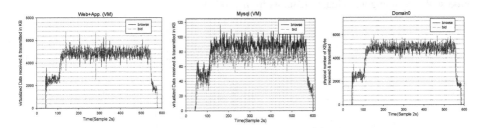

**Fig. 4.** Network data received and transmitted by the web and application servers and the database servers in VMs and the hypervisor.

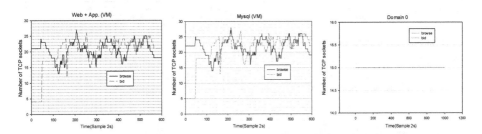

**Fig. 5.** Number of TCP Sockets generated by the web and application servers and the database servers in VMs and the hypervisor.

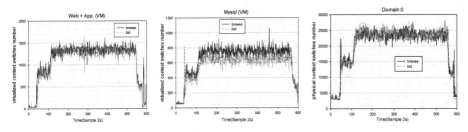

**Fig. 6.** Number of context switches performed by the web and application servers and the database servers in VMs and the hypervisor.

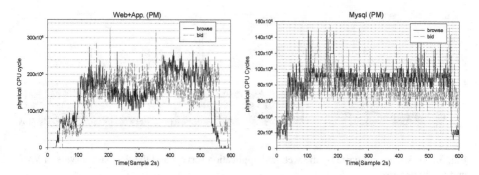

**Fig. 7.** CPU cycle demands by the web and application servers and the database servers to process the browsing and bidding requests from 1000 clients.

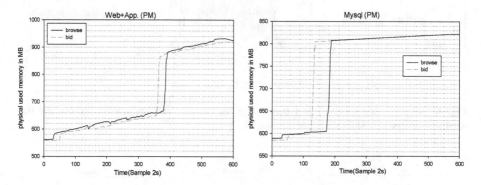

**Fig. 8.** RAM demands by the web and application servers and the database servers.

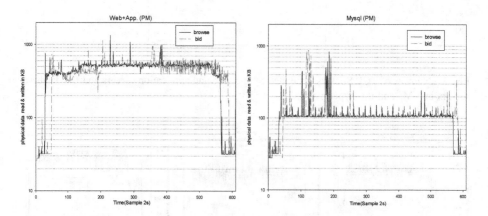

**Fig. 9.** Disk read and write by the web and application servers and the database servers.

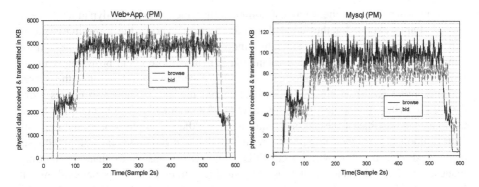

**Fig. 10.** Network data received and transmitted by the web and application servers and the database servers.

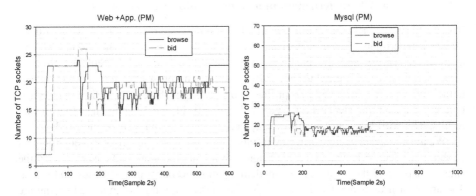

**Fig. 11.** Number of TCP Sockets created by the web and application servers and the database servers to process the browsing and bidding requests from 1000 clients.

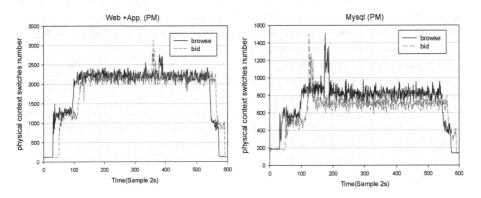

**Fig. 12.** Number of context switches performed by the web and application servers and the database servers.

Comparing the two client request compositions, their workload dynamics display similar shapes except for the RAM demands. Figure 2 shows the browsing requests experience one or more jumps demanding more RAM, while the bidding requests have a more smooth curve. A possible explanation is that as more client browsing requests arrive, some requests are backlogged and after a certain period of time the server allocates more RAM to process those backlogged requests, which also causes more disk reads/writes (the spikes in the first two sub-figures of Fig. 3). On the other hand, the longer think time of the bidding requests allows the servers to process the requests more smoothly. Another important finding is that although the browsing requests demand similar or more virtualized CPU and network resources than the bidding requests, the latter demands a little more physical resources than the former as shown in Figs. 1 and 4. We also find that the number of context switches performed by the hypervisor is larger than that by VMs servers as shown in Fig. 6.

### 4.2    Workload Characterization in a Non-virtualized Environment

In order to characterize the impact of virtualization on system's workload, we conduct a series of experiment on non-virtualized servers in our testbed. The front-end web and application servers and the back-end database servers reside on separate physical servers. 1000 clients external to the RUBiS servers send browsing and bidding requests to the web server. *Sysstat* and *perf* profile resource usages directly from the host OS and hardware on each physical server. Figures 7, 8, 9, 10, 11 and 12 show the experimental results. The workload curves still display certain patterns that can be modeled.

We are interested in comparing the results with those from the virtualized environment as shown in Sect. 4.1. The two sets of figure show that the workload curves display the similar shapes and the front-end servers demand more resource than the back-end server. The aggregated demands for the six types of resources in the non-virtualized setting are 3.47, 0.97, 0.6, 0.98, 2.67 and 0.12 times more/less than those in the virtualized environment. The workload requests for RAM show the most significant difference between the two environments. As in the non-virtualized system (Fig. 8), the bidding requests also display abrupt increase of RAM usage and the jumps happen earlier in time than those in the virtualized system. One reason is the longer communication delay in the non-virtualized system. In addition, from Fig. 9 we can see disk read and write workload shows higher variance in the non-virtualized system than the virtualized one.

Comparing the results in Sects. 4.1 and 4.2, we find application's demand for physical resources is higher in the non-virtualized environment than in the virtualized one, with 88 % more CPU cycles, 21 % more RAM, 2 % more network traffic and 60 % more TCP sockets, while disk read/write is 25 % less and the number of context switches is 87 % less. These findings will allow cloud service providers to achieve efficient capacity planning for a desirable SLA satisfaction rate.

## 5    Conclusion

In this work, we study the impact of virtualization on the workload dynamics. We present experimental results on a cloud testbed by profiling the workload dynamics on both virtualized and non-virtualized environments. We compare the resource demands at the three server tiers.

This study is preliminary. Our goal is to extract the rules of thumb to aid cloud service providers to achieve the best resource planning. We plan to design and apply formal methods to model the workload dynamics at both resource level and transaction level.

**Acknowledgment.** We would like to thank the anonymous reviewers for their constructive comments and suggestions. A preliminary version of this paper was accepted by the fourth Workshop on Big Data Benchmarks, Performance Optimization and Emerging Hardware in conjunction with ACM ASPLOS 2014 [27]. This work was performed in the Dependable Computing Systems Laboratory at the University of North Texas.

# References

1. Armbrust, M., Fox, A., Griffith, R., Joseph, A.D., Katz, R., Konwinski, A., Lee, G., Patterson, D., Rabkin, A., Stoica, I., Zaharia, M.: A view of cloud computing. Commun. ACM **53**(4), 50–58 (2010)
2. Smith, J.E., Nair, R.: The architecture of virtual machines. IEEE Comput. **38**(5), 32–38 (2005)
3. Wang, Z., Tang, X., Luo, X.: Policy-based SLA-aware cloud service provision framework. In: Proceedings of IEEE International Conference on Semantics Knowledge and Grid (SKG) (2011)
4. Guan, Q., Chiu, C., Fu, S.: CDA: a cloud dependability analysis framework for characterizing system dependability in cloud computing infrastructures. In: Proceedings of IEEE the 18th International Symposium on Dependable Computing (PRDC) (2012)
5. Wang, G., Eugene Ng, T.S.: The impact of virtualization on network performance of Amazon EC2 data center. In: Proceedings of IEEE Conference on Computer Communications (INFOCOM) (2010)
6. RUBiS Website. http://rubis.ow2.org
7. Hernández-Orallo, E., Vila-Carb, J.: Web server performance analysis using histogram workload models. Comput. Netw. **53**(15), 2727–2739 (2009)
8. Shi, W., Wright, Y., Collins, E., Karamcheti, V.: Workload characterization of a personalized web site and its implications for dynamic content caching. In: Proceedings of International Conference on Web Content Caching and Distribution (WCW) (2002)
9. Thereska, E., Donnelly, A., Narayanan, D.: Sierra: practical powerproportionality for data center storage. In: Proceedings of ACM European Conference on Computer Systems (EuroSys) (2011)
10. Bairavasundaram, L.N., Arpaci-Dusseau, A.C., Arpaci-Dusseau, R.H., Goodson, G.R., Schroeder, B.: An analysis of data corruption in the storage stack. ACM Trans. Storage **4**(3), 821–834 (2008)
11. Wang, F., Xin, Q., Hong, B., Brandt, S.A., Miller, E.L., Long, D.D.E., Mclarty, T.T.: File system workload analysis for large scale scientific computing applications. In: Proceedings of IEEE Conference on Mass Storage Systems and Technologies (MSST) (2004)
12. Ersoz, D., Yousif, M.S., Das, C.R.: Characterizing network traffic in a cluster-based, multi-tier data center. In: Proceedings of IEEE International Conference on Distributed Computing Systems (ICDCS) (2007)

13. Paxson, V.: Empirically derived analytic models of wide-area TCP connections. IEEE/ACM Trans. Netw. **2**(4), 316–336 (1994)
14. Christodoulopoulos, K., Gkamas, V., Varvarigos, E.A.: Statistical analysis and modeling of jobs in a grid environment. J. Grid Comput. **6**(1), 77–101 (2008)
15. Medernach, E.: Workload analysis of a cluster in a grid environment. In: Feitelson, D.G., Frachtenberg, E., Rudolph, L., Schwiegelshohn, U. (eds.) JSSPP 2005. LNCS, vol. 3834, pp. 36–61. Springer, Heidelberg (2005)
16. Song, B., Ernemann, C., Yahyapour, R.: User group-based workload analysis and modelling. In: Proceedings of IEEE/ACM International Symposium on Cluster, Cloud and Grid Computing (CCGrid) (2005)
17. Iamnitchi, A., Doraimani, S., Garzoglio, G.: Workload characterization in a high-energy data grid and impact on resource management. Clust. Comput. **12**(2), 153–173 (2009)
18. Wang, L., Zhan, J., Luo, C., Zhu, Y., Yang, Q., He, Y., Gao, W., Jia, Z., Shi, Y., Zhang, S., Zheng, C., Lu, G., Zhan, K., Li, X., Qiu, B.: BigDataBench: a big data benchmark suite from Internet services. In: Proceedings of IEEE International Symposium on High Performance Computer Architecture (HPCA) (2014)
19. Luo, C., Zhan, J., Jia, Z., Wang, L., Lu, G., Zhang, L., Xu, C., Sun, N.: CloudRank-D: benchmarking and ranking cloud computing systems for data processing applications. Front. Comput. Sci. **6**(4), 347–362 (2012)
20. D'Ambrogio, A., Bocciarelli, P.: A model-driven approach to describe and predict the performance of composite services. In: Proceedings of ACM International Workshop on Software and Performance (WOSP) (2007)
21. Kavulya, S., Tan, J., Gandhi, R., Narasimhan, P.: An analysis of traces from a production mapreduce cluster. In: Proceedings of IEEE/ACM International Symposium on Cluster, Cloud and Grid Computing (CCGrid) (2010)
22. Mishra, A.K., Hellerstein, J.L., Cirne, W., Das, C.R.: Towards characterizing cloud backend workloads: insights from google compute clusters. SIGMETRICS Perform. Eval. Rev. **37**(4), 34–41 (2010)
23. Cooper, B.F., Silberstein, A., Tam, E., Ramakrishnan, R., Sears, R.: Benchmarking cloud serving systems with YCSB. In: Proceedings of ACM Symposium on Cloud Computing (SOCC) (2010)
24. Wang, L., Tao, J., Kunze, M., Castellanos, A.C., Kramer, D., Karl, W.: Scientific cloud computing: early definition and experience. In: Proceedings of IEEE International Conference on High Performance Computing and Communications (HPCC) (2008)
25. Sysstat. http://sebastien.godard.pagesperso-orange.fr
26. Perf. http://en.wikipedia.org/wiki/Perf_(Linux)
27. Wang, X., Huang, S., Fu, S., Kavi, K.: Characterizing workload of web applications on virtualized servers. In: Accepted by Workshop on Big Data Benchmarks, Performance Optimization, and Emerging Hardware (BPOE) in conjunction with the 19th ACM International Conference on Architectural Support for Programming Languages and Operating Systems (ASPLOS) (2014)

# Topical Section Headings: Performance Optimization and Evaluation

# Performance Benefits of DataMPI: A Case Study with BigDataBench

Fan Liang[1,2(✉)], Chen Feng[1,2], Xiaoyi Lu[3], and Zhiwei Xu[1]

[1] Institute of Computing Technology, Chinese Academy of Sciences, Beijing, China
{liangfan,fengchen,zxu}@ict.ac.cn
[2] University of Chinese Academy of Sciences, Beijing, China
[3] Department of Computer Science and Engineering,
The Ohio State University, Columbus, USA
luxi@cse.ohio-state.edu

**Abstract.** Apache Hadoop and Spark are gaining prominence in Big Data processing and analytics. Both of them are widely deployed in Internet companies. On the other hand, high-performance data analysis requirements are causing academical and industrial communities to adopt state-of-the-art technologies in HPC to solve Big Data problems. Recently, we have proposed a key-value pair based communication library, DataMPI, which is extending MPI to support Hadoop/Spark-like Big Data Computing jobs. In this paper, we use BigDataBench, a Big Data benchmark suite, to do comprehensive studies on performance and resource utilization characterizations of Hadoop, Spark and DataMPI. From our experiments, we observe that the job execution time of DataMPI has up to 57 % and 50 % speedups compared with those of Hadoop and Spark, respectively. Most of the benefits come from the high-efficiency communication mechanisms in DataMPI. We also notice that the resource (CPU, memory, disk and network I/O) utilizations of DataMPI are also more efficient than those of the other two frameworks.

**Keywords:** DataMPI · Hadoop · Spark · BigDataBench

## 1 Introduction

Data explosion is becoming an irresistible trend with the development of Internet, social network, e-commerce, etc. Over the last decade, there have been emerging a lot of systems and frameworks for Big Data, such as Hadoop [1], Dyrad [8], Yahoo! S4 [14] and so on. Apache Hadoop has become as the defacto standard for Big Data processing and analytics. Many clusters in the production environment already contain thousands of nodes to dedicatedly run Hadoop jobs everyday. Beyond the success of Hadoop, Spark [19] provides another feasible way to process large amount of data by introducing the in-memory computing techniques. Nowadays, both of them have attracted more and more attention from academical and industrial areas.

© Springer International Publishing Switzerland 2014
J. Zhan et al. (Eds.): BPOE 2014, LNCS 8807, pp. 111–123, 2014.
DOI: 10.1007/978-3-319-13021-7_9

However, the performance of current commonly used Big Data systems is still in a sub-optimal level. Many studies [9,11,15,17] have been trying to adopt state-of-the-art technologies in the High Performance Computing (HPC) area to accelerate the performance of Big Data processing. As one example of these attempts, our previous work [12,13,18] shows the performance of Hadoop communication primitives still have huge performance improvement potentials, and Message Passing Interface (MPI), which is widely used in the field of HPC, can help to optimize communication performance of Hadoop. Furthermore, the key-value pair based communication library, DataMPI [3,12], has been proposed to efficiently execute Hadoop/Spark-like Big Data Computing jobs by extending MPI. Since the open-source nature of these systems, it will be very interesting for users to know the performance characteristics of the emerging Big Data systems by doing a systematical performance evaluation over different aspects.

In this paper, we use BigDataBench [16], one of the benchmark suites for Big Data Computing systems, to evaluate the performance of Hadoop, Spark and DataMPI. By tracing the resource utilization, we analyse the execution behavior of each system. Our contributions in this paper include

- We propose a seven-pronged approach to evaluate Big Data Computing systems, which can help researchers to understand the performance of those systems systematically.
- Evaluation results show DataMPI can achieve up to 57 % and 50 % speedups compared to Hadoop and Spark, respectively, for the high-efficiency communication mechanisms and lightweight software design.

The rest of this paper is organized as follows. Section 2 discusses background and related work. Section 3 states our experiments methodology. The evaluation results and analysis are given in Sect. 4. Section 5 concludes the paper.

## 2    Background and Related Work

### 2.1    Big Data Systems

MapReduce programming model is pivotal in Big Data Computing. Hadoop [1], one of the open-source implementations of MapReduce, is becoming the defacto standard. It has been widely used in various areas and applications, such as log analysis, machine learning, search engine, etc., and achieves success for its high scalability, built-in fault-tolerance and simplicity of programming. Spark [19], one of the emerging Big Data Computing systems, processes task-parallel jobs with in-memory techniques. It implements resilient distributed datasets (RDDs), the distributed memory abstraction which builds on the lineage concept, and performs efficiently in iterative algorithms of machine learning and interactive data mining. DataMPI [12] is a key-value pair based communication library which extends MPI for Hadoop/Spark-like Big Data Computing systems. It implements a bipartite communication model and leverages the state-of-the-art technologies of MPI in HPC to accelerate the execution performance of Big Data applications.

Those three Big Data systems are respectively typical for the different implementation technologies based on their particular execution models. This motivates us to comprehensively evaluate them with Big Data benchmarks. Besides, many other Big Data systems using HPC technologies have been emerging. The authors in [9,11,15] implement Hadoop-RDMA, which uses RDMA-capable (Remote Direct Memory Access) interconnects to enhance the design of Hadoop. Wang et al. [17] propose the network-levitated merge algorithm and implement Hadoop-A which overlaps data merge and reduce operations for Hadoop Reduce tasks.

### 2.2   Big Data Benchmarks

Researchers have proposed several benchmarks for evaluating Big Data Computing systems. MRBench [10] is designed for evaluating MapReduce frameworks using TPC-H workloads. HiBench [7] is designed for Hadoop-based systems based on micro-benchmarks, web search, machine learning and HDFS benchmarks. BigBench [6], an end-to-end benchmark proposal based on product retailer, is designed for parallel DBMS and MapReduce systems. BigDataBench [16] is a benchmark suite for different Big Data Computing systems, such as Hadoop, Spark, etc. It covers six typical application scenarios which include fundamental workloads and Internet service applications. BigDataBench also provides a data generator, Big Data Generator Suite (BDGS) which extracts the characteristics of real-world data, to create synthetic data sets.

As BigDataBench contains various workloads, and synthetic data generator, we choose it as our benchmark suite. According to the specification of BigDataBench, we use DataMPI to implement the benchmarks and evaluate the performance of Hadoop, Spark and DataMPI fairly.

## 3   Benchmarking Methodology

### 3.1   Chosen Workloads

We choose five typical workloads in BigDataBench as our benchmarks, which include three micro-benchmarks and two application benchmarks.

- **Micro-benchmarks** include *WordCount*, *Grep* and *Sort*, which are fundamental and widely used operations in broad analysis processes. *WordCount* counts the number of each word occurrence in a collection of documents. *Grep* searches strings conforming to a certain pattern in the input documents and counts the number of the occurrence of each matched string. *Sort* reads each record of the input files as a key-value pair and sorts the records based on the keys.
- **Application benchmarks** include *K-means* and *Naive Bayes*, which are typical applications in social network and e-commerce scenarios. *K-means* is a classical clustering algorithm in data mining which aims to partition the input objects to $k$ clusters by calculating the nearest mean cluster of each

object belongs to. *Naive Bayes* is a probabilistic algorithm for classification. It is based on Bayes' theorem with strong independence assumption, which means the features of the model are independent with each other.

## 3.2  Evaluation Methodology

We record the execution time of each workload over one system, which reflects the system performance. To understand the runtime status, we monitor the systems with resource metrics which include:

- CPU utilization: it is recorded as a percentage of CPU usage and shows the time the total CPU has spent on running a workload. The CPU utilization will be recorded each second. We calculate the average CPU utilization during one workload execution.
- Network I/O throughput: it is defined as the average amount of data transmitted (send/receive) per second over the network.
- Disk I/O throughput: it is defined as the average amount of data transmitted (read/write) per second through the hard disks.
- Memory footprint: it refers to the memory used when running a workload. The memory footprint will dynamically change when system allocates and releases memory during workload execution. We calculate the average memory footprint during one workload execution to compare memory utilization.

To evaluate Hadoop, Spark and DataMPI, we follow a seven-pronged approach as shown in Fig. 1. To show the performance, we calculate the average execution time of each kind of workloads over each system. The small job is based on the micro-benchmarks, while data size of each workload is 128 MB. The data sizes of normal micro-benchmarks and application benchmarks vary

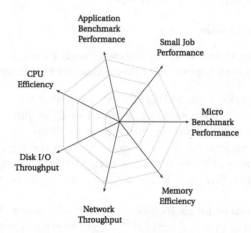

**Fig. 1.** Evaluation methodology

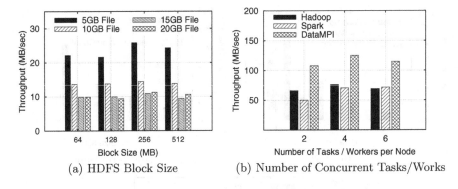

**Fig. 2.** Parameter tuning

from 8 GB to 64 GB. The CPU efficiency is defined as the CPU usage for the workload. The memory efficiency reflects the average amount of memory allocated to the system. The less memory is used, the better memory efficiency is achieved. We summarize the results with these seven dimensions in Sect. 4.7. To better understand the results, we use the execution time and resource utilization of Hadoop as the baseline and normalize the corresponding values of Spark and DataMPI.

# 4   Experimental Evaluation

## 4.1   Experiment Setup

We use a cluster composed of 8 nodes interconnected by a 1 Gigabit Ethernet switch as our testbed. Each node is equipped with two Intel Xeon E5620 CPU processors (2.4 GHz) with disabling the hyper-threads. Each processor has four physical cores. Each node has 16 GB DDR3 RAM with 1333 MHz and one 150 GB free space SATA disk.

The operation system used is CentOS release 6.5 (Final) with kernel version 2.6.32-431.el6.x86_64. The software stack is comprised with JDK 1.7.0_25, MVAPICH2-2.0b, Scala 2.9.3, BigDataBench 2.1, Hadoop 1.2.1, Mahout 0.8 [2], Spark 0.8.1 and DataMPI 0.6. For all evaluations, we report results that are averaged across three executions.

## 4.2   Chosen Parameters

Hadoop, Spark and DataMPI have abundant parameters to set to achieve better performance. In this section, we tune the parameters for fair evaluations. We mainly focus on the HDFS block size and the number of tasks or workers, because the disk and network will easily become the bottlenecks in our testbed, and the concurrent execution instances have a great influence on performance.

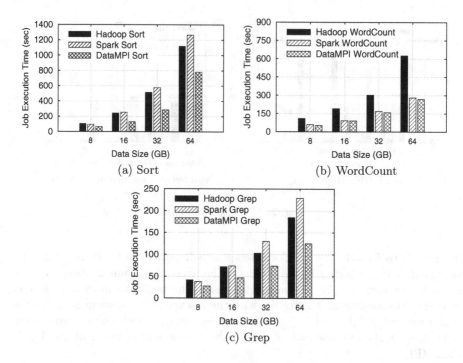

**Fig. 3.** Performance comparison of different micro-benchmarks

We use DFSIO program, a file system level benchmark of Hadoop, as the workload for tuning HDFS block size. We measure the throughput by varying the HDFS block size and input data size. Figure 2(a) shows the throughput achieves the best, when block size is 256 MB. When tuning the number of concurrent tasks or workers, we execute Sort benchmark and measure the throughput by processing 1 GB data per Hadoop/DataMPI task and Spark worker with increasing the number of concurrent tasks or workers from 2 to 6 per node. Figure 2(b) shows the systems can get the best throughput when the number of concurrent tasks or workers on each node is 4.

Based on the two tests, we run our following evaluations based on 256 MB HDFS block size and 4 concurrent tasks or workers per node. The replication of each block in HDFS is set to three for the high data availability and flexible data locality.

## 4.3 Micro-benchmark Performance

In this section, we evaluate the performance of micro-benchmarks among Hadoop, Spark and DataMPI. We use BDGS in BigDataBench to produce the data sets and upload them on the HDFS with the uncompressed text format. The seed model used in BDGS is *lda_wiki1w* which is trained from wikipedia entries corpus.

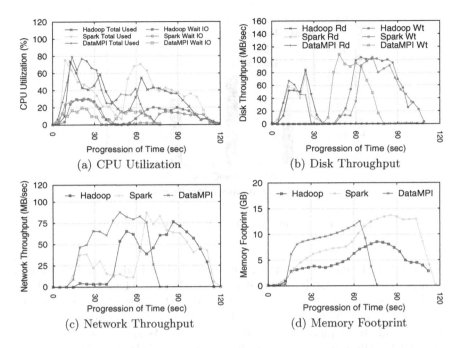

**Fig. 4.** Resource utilization of Sort benchmark with 8 GB data

We vary the input data size from 8 GB to 64 GB. Figure 3(a) shows DataMPI has 30 %–44 % (averagely 39 %) improvement compared to Hadoop and 38 %–50 % (averagely 44 %) improvement compared to Spark, when running Sort. Figure 3(b) shows DataMPI and Spark have similar performance and achieve 47 %–57 % (averagely 52 %) performance improvement compared to Hadoop, when running WordCount. The Grep evaluation results in Fig. 3(c) show that DataMPI cuts down the execution time by 29 %–34 % (averagely 32 %) compared to Hadoop, and by 26 %–45 % (averagely 38 %) compared to Spark.

DataMPI can achieve the best performance for those benchmarks, while Spark is not performing better than Hadoop in Sort and Grep cases, which means for batch jobs, Hadoop is still relatively good.

### 4.4   Profile of Resource Utilization

We profile the resource utilization of Hadoop, Spark and DataMPI based on the workloads of 8 GB Sort and 32 GB WordCount from four aspects, i.e. CPU utilization, disk throughput, network throughput and memory footprint. We record the total CPU usage percentage and the CPU wait I/O percentage. A higher CPU wait I/O percentage means CPU costs more time to wait for I/O operations to complete. For the sake of page limitation, we only show the figures of Sort benchmark with 8 GB case.

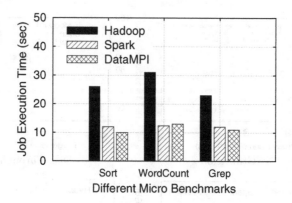

**Fig. 5.** Performance comparison based on small jobs

For the Sort benchmark, DataMPI costs 69 s while Hadoop and Spark cost 117 s and 114 s, respectively. Hadoop has Map/Reduce phases, DataMPI has O/A phases, and Spark has map phase (Stage0), sort-by-key phase (Stage1) and save phase (Stage2). The Map phase of Hadoop costs 36 s, the Stage 0 of Spark costs 38 s, and the O phase of DataMPI costs 28 s. As shown in Fig. 4(a), the average CPU utilizations of Hadoop, Spark and DataMPI are 37 %, 38 % and 45 %. The average CPU wait I/O percentages of Hadoop, Spark and DataMPI are 15 %, 12 % and 10 %. This means Hadoop costs more time to wait I/O operations. Figure 4(b) shows the disk throughput. The average disk read throughputs of Hadoop Map phase, Spark Stage 0 and DataMPI O phase are 49 MB/s, 46 MB/s and 50 MB/s. The average disk write throughputs of Hadoop Shuffle-Reduce phase, Spark Stage 2 and DataMPI A phase are 67 MB/s, 66 MB/s and 69 MB/s. Figure 4(c) shows the network throughput of DataMPI is averagely 62 MB/s, which is 59 % higher than that of Hadoop (39 MB/s) and 55 % higher than that of Spark (40 MB/s). This means MPI-based communication mechanism can use network resource more efficiently. Figure 4(d) shows the average memory footprints of Hadoop, Spark and DataMPI are 5 GB, 9 GB and 8 GB.

When running WordCount, Spark and DataMPI cost 169 s and 158 s, and have 47 %, 43 % speedups compared to Hadoop (301 s), respectively. The CPU utilizations of Hadoop, Spark and DataMPI are 82 %, 58 % and 84 %, respectively. The average read throughput of DataMPI is 44 MB/s, which is approximately equal to that of Spark and is higher than that of Hadoop (20 MB/s). We observe that DataMPI and Hadoop have few network transmissions, while Spark transmits more intermediate data for the RDDs creation. The average memory footprints of Hadoop, Spark and DataMPI are 8 GB, 6 GB, and 5 GB.

From the above two cases, we observe that DataMPI can leverage the resources to run jobs more effectively than Hadoop and Spark for the high efficient data communication and computation.

## 4.5  Small Jobs

According to the recent study [5], more than 90 % of MapReduce jobs in Facebook and Yahoo! are small jobs, which means the input data sizes of the jobs are usually kilo or mega bytes. The system overheads of the initialization and the finalization have serious impacts on the performance. In this section, we compare the performance of Hadoop, Spark and DataMPI when they run the micro-benchmarks of Sort, WordCount and Grep with smaller input data sets. The input data size of each workload is 128 MB. The number of the concurrent tasks or works is set to one per node. Figure 5 shows that DataMPI has similar performance with Spark, and performs averagely 54 % more efficiently than Hadoop. The benefits of Spark and DataMPI are contributed by the lightweight software designs.

## 4.6  Application Benchmark Performance

In this section, we present the results of the application evaluations. The Hadoop implementations of K-means and Naive Bayes in BigDataBench are based on Mahout, while the Spark implementation of K-means is based on Spark MLlib [4]. We implement K-means and Naive Bayes over DataMPI according to the Big-DataBench specification. Because BigDataBench 2.1 lacks the implementation of Naive Bayes in Spark, we only compare the performance of this benchmark between DataMPI and Hadoop. We first explain the processing characteristics of the applications from the implementation-level and then give the performance results.

**K-means:** We use BDGS to generate the input data sets based on five seed models, amazon1-amazon5. Using *genData_Kmeans* of BDGS, text files are converted to sequence files from directories, then transformed to the sparse vectors as the input data of training clusters. Our evaluations are based on the sparse vectors and mainly focus on the performance of training execution. As stated in Sect. 3.1, K-means trains the cluster centroids iteratively. Each iteration of K-means is a MapReduce job. In one job, Map tasks read the initial or previous cluster centroids from HDFS, afterwards, assign the input vectors to appropriate clusters according to the distance calculation and train the new centroids independently. At the end of Map tasks, new centroids will be sent to the Reduce tasks according to the cluster indexes. Reduce tasks receive and update the centroids for the next iterative execution. We observe that most of K-means calculation happens in Map phase, and few intermediate data is generated.

Our tests show Spark has outstanding performance when running the iterative computations based on the RDDs. Since Hadoop is not designed for iterative jobs, for fair comparison, we record the execution time of the first iteration from the job start, which considering the overheads of loading data, computation and communication, and outputing results. Figure 6(a) shows that DataMPI has at most 39 % improvement than Hadoop and 33 % improvement than Spark when the input data size varies from 8 GB to 64 GB.

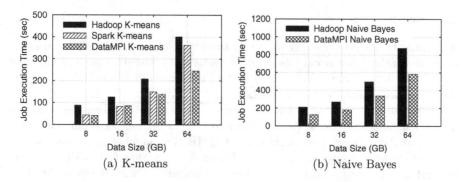

**Fig. 6.** Performance comparison of application benchmarks

**Naive Bayes:** The input document data sets are generated by BDGS, and are classified into five categories. The procedure of Naive Bayes mainly contains two steps, including converting sequence files to sparse vectors and training the Naive Bayes model. Mahout runs several MapReduce jobs to create the sparse vectors. Firstly, one document is converted to a token array. After that, some MapReduce jobs are launched to count the term frequency in one document and document frequency of all terms. The sparse vector of one document is calculated according to the term frequency and document frequency. The main operation in above steps is counting, including term counting and document counting, which means that the behavior of Naive Bayes is similar to WordCount. In our evaluation cases, the data sizes of sparse vector and term-counting dictionary are within several mega bytes. The model training contains two MapReduce jobs to execute the probabilistic computations. The two jobs cost less time than the sparse vectors generation because of the simple calculating operations and small input data sizes. Figure 6(b) shows DataMPI has 33 % improvement than Hadoop averagely.

### 4.7 Discussion of Performance Results

We summarize the performance comparisons with different benchmarks using seven-pronged diagram, depicted in Fig. 7. We normalize the values of Spark and DataMPI according to the corresponding Hadoop values. Besides, we only take the K-means results to calculate the values of the application benchmarks. Compared to Hadoop, DataMPI can averagely achieve 41 %, 54 % and 38 % performance improvements when running micro-benchmarks, small jobs and application benchmarks, respectively, while Spark has 10 %, 54 % and 31 % performance improvements, respectively. From the Sort and WordCount cases, the average CPU utilizations of Hadoop, Spark and DataMPI are 60 %, 48 % and 64 %, which means DataMPI has similar CPU efficiency with Hadoop, and leverages the CPU resource 33 % more efficiently than Spark. The average disk I/O throughputs of Spark and DataMPI are 15 %, 20 % higher than that of Hadoop, respectively. DataMPI achieves 56 % and 55 % network throughput

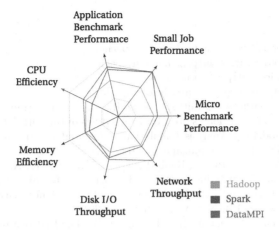

**Fig. 7.** Evaluation results

improvements than those of Spark and Hadoop, respectively. The average memory footprints of Hadoop, Spark and DataMPI are 6.5 GB, 7.5 GB and 6.5 GB, which means Hadoop and DataMPI can efficiently utilize memory when running the workloads. The benefits of DataMPI come from the lightweight software design and the high performance communication design which is able to leverage system resources to pipeline the computation and communication operations efficiently [12].

## 5   Conclusion

In this paper, we provide a systematical performance evaluation of Hadoop, Spark and DataMPI based on BigDataBench. We choose three micro benchmarks (Sort, WordCount and Grep) and two application benchmarks (K-means and Naive Bayes) as our testing experimental workloads. Based on the Sort, WordCount benchmark cases, we present a detailed resource utilization analysis of the three systems. Our evaluation shows that with the mirco-benchmarks, DataMPI can achieve 29 %–57 % performance improvement compared to Hadoop, and up to 50 % performance improvement compared to Spark. The small job evalutions show the low overheads of DataMPI and Spark make them gain 54 % performance improvement compared to Hadoop. Evaluations of K-means and Naive Bayes benchmarks show DataMPI can achieve 33 %–39 % application-level performance improvement compared to Hadoop and Spark.

**Acknowledgments.** This work is supported in part by the Strategic Priority Program of Chinese Academy of Sciences (Grant No. XDA06010401). We thank the anonymous referees for their comments. We also thank Prof. Li Zha for his help to support this research.

# References

1. Apache Hadoop Project. http://hadoop.apache.org
2. Apache Mahout Project. https://mahout.apache.org
3. DataMPI Project. http://datampi.org
4. MLlib Project. https://spark.apache.org/mllib
5. Chen, Y., Ganapathi, A., Griffith, R., Katz, R.: The case for evaluating MapReduce performance using workload suites. In: Proceedings of the 19th International Symposium on Modeling, Analysis Simulation of Computer and Telecommunication Systems (MASCOTS '11), Singapore (2011)
6. Ghazal, A., Rabl, T., Hu, M., Raab, F., Poess, M., Crolotte, A., Jacobsen, H.A.: BigBench: towards an industry standard benchmark for big data analytics. In: Proceedings of the 32nd ACM SIGMOD International Conference on Management of Data (SIGMOD '13), New York, NY, USA (2013)
7. Huang, S., Huang, J., Dai, J., Xie, T., Huang, B.: The HiBench benchmark suite: characterization of the MapReduce-based data analysis. In: Proceedings of the 26th International Conference on Data Engineering Workshops (ICDEW '10), Long Beach, CA, USA (2010)
8. Isard, M., Budiu, M., Yu, Y., Birrell, A., Fetterly, D.: Dryad: distributed data-parallel programs from sequential building blocks. ACM SIGOPS Oper. Syst. Rev. **41**(3), 59–72 (2007)
9. Islam, N.S., Rahman, M.W., Jose, J., Rajachandrasekar, R., Wang, H., Subramoni, H., Murthy, C., Panda, D.K.: High performance RDMA-based design of HDFS over InfiniBand. In: Proceedings of the 25th International Conference on High Performance Computing, Networking, Storage and Analysis (SC '12), Salt Lake City, UT, USA (2012)
10. Kim, K., Jeon, K., Han, H., Kim, S., Jung, H., Yeom, H.: MRBench: a benchmark for MapReduce framework. In: Proceedings of the 14th International Conference on Parallel and Distributed Systems (ICPADS '08), Melbourne, Victoria, Australia (2008)
11. Lu, X., Islam, N.S., Wasi-ur-Rahman, M., Jose, J., Subramoni, H., Wang, H., Panda, D.K.: High-performance design of Hadoop RPC with RDMA over Infini-Band. In: Proceedings of the 42nd International Conference on Parallel Processing (ICPP '13), Lyon, France (2013)
12. Lu, X., Liang, F., Wang, B., Zha, L., Xu, Z.: DataMPI: extending MPI to Hadoop-like big data computing. In: Proceedings of the 28th International Parallel and Distributed Processing Symposium (IPDPS '14), Phoenix, AZ, USA (2014)
13. Lu, X., Wang, B., Zha, L., Xu, Z.: Can MPI benefit Hadoop and MapReduce applications? In: Proceedings of the 40th International Conference on Parallel Processing Workshops (ICPPW '11), Taipei, China (2011)
14. Neumeyer, L., Robbins, B., Nair, A., Kesari, A.: S4: distributed stream computing platform. In: Proceedings of the 10th IEEE International Conference on Data Mining Workshops (ICDMW '10), Sydney, Australia (2010)
15. Wasi-ur Rahman, M., Islam, N., Lu, X., Jose, J., Subramoni, H., Wang, H., Panda, D.: High-performance RDMA-based design of Hadoop MapReduce over InfiniBand. In: Proceedings of the 27th International Symposium on Parallel and Distributed Processing Workshops and PhD Forum (IPDPSW '13), Cambridge, MA, USA (2013)

16. Wang, L., Zhan, J., Luo, C., Zhu, Y., Yang, Q., He, Y., Gao, W., Jia, Z., Shi, Y., Zhang, S., Zheng, C., Lu, G., Zhan, K., Li, X., Qiu, B.: BigDataBench: a big data benchmark suite from internet services. In: Proceedings of the 20th IEEE International Symposium on High Performance Computer Architecture (HPCA '14), Orlando, FL, USA (2014)

17. Wang, Y., Que, X., Yu, W., Goldenberg, D., Sehgal, D.: Hadoop acceleration through network levitated merge. In: Proceedings of the 24th International Conference for High Performance Computing, Networking, Storage and Analysis (SC '11), New York, NY, USA (2011)

18. Xu, Z.: High-performance techniques for big data computing in internet services. In: Proceeding of the 2012 SC Companion: High Performance Computing (SC '12), Salt Lake City, UT, USA (2012)

19. Zaharia, M., Chowdhury, M., Das, T., Dave, A., Ma, J., McCauley, M., Franklin, M., Shenker, S., Stoica, I.: Resilient distributed datasets: a fault-tolerant abstraction for in-memory cluster computing. In: Proceedings of the 9th USENIX Symposium on Networked Systems Design and Implementation (NSDI '12), San Jose, CA, USA (2012)

# InvarNet-X: A Comprehensive Invariant Based Approach for Performance Diagnosis in Big Data Platform

Pengfei Chen[✉], Yong Qi, Di Hou, and Huachong Sun

School of Electronic and Information Engineering, Xi'an Jiaotong University,
Xianning West Road No.28, Xi'an 710049, China
chenpengfei@outlook.com,
{qiy,dihou}@mail.xjtu.edu.cn,
huachong612@gmail.com

**Abstract.** To provide a high performance and reliable big data platform, this paper proposes a comprehensive invariant-based performance diagnosis approach named *InvarNet-X*. *InvarNet-X* not only covers performance anomaly detection but also root cause inference, both of which are conducted under the consideration of operation context of big data applications. The performance anomaly detection procedure is adopted to trigger the cause inference procedure and accomplished by checking the ARIMA model drift on Cycle Per Instruction (CPI) data of big data applications. The oracle of cause inference is the unobservable root causes of performance problems always expose themselves via the violations of the associations amongst directly observable performance metrics. In *InvarNet-X*, such observable associations as the likely invariants are established by the Maximal Information Criteria (MIC) and each performance problem is signified by a set of violations of those likely invariants. Finally, the root cause is uncovered by searching a similar signature in the signature database. With such a comprehensive analysis, *InvarNet-X* can provide much detailed clues for performance problems and even pinpoint the root causes if the signature database is given. Through experimental evaluations in a small prototype, we find out *InvarNet-X* can achieve an average 91 % precision and 87 % recall in diagnosing some real faults reported in software bug repositories, which is superior to several state-of-the-art approaches. Meanwhile, the local modeling methodology makes *InvarNet-X* easily facilitated in real-time and large scale big data platforms.

**Keywords:** Big data · Hadoop · Observable likely invariant · Performance diagnosis

## 1 Introduction

Big data becomes an inevitable trend at present and in the foreseeable future. The popularity of big data attracts many researchers and engineers to devote

© Springer International Publishing Switzerland 2014
J. Zhan et al. (Eds.): BPOE 2014, LNCS 8807, pp. 124–140, 2014.
DOI: 10.1007/978-3-319-13021-7_10

themselves to mining the valuable knowledge in the scrambled data piles. However, during the transformation from 'big data' to 'big value', the performance and reliability of the big data platform deserves the same attention. As a general case, the big data platforms, most if not all, are deployed in large scale distributed systems with thousands of machines using parallel programming such as MapReduce as their program paradigm. In such a huge platform, performance anomalies, faults and failures become commonplace due to the complex interactions in the intricate software stacks [1]. In our previous study [2], we summarized the causes of faults in several widely used open software systems such as Hadoop. One part of the causes are the operational environment changes such as resource utilization hog, workload fluctuation and misconfiguration and the other part are the bugs rooted in the software stacks such as memory leak and lock race. In the big data software stacks, hadoop, no-sql databases, et al. are all the candidate hotbeds of these faults. In addition to that, new faults emerge in the big data platform due to the inherent complexity and three "V"s (i.e. Velocity, Volume and Variety[1]) of big data. The typical bugs are out of memory (OOM) and disk space exhaustion. For instance a bug, MapReduce-1182, shows OOM when the data becomes huge. The bug tells us under the low data-intensive workloads, "shuffle" in memory may be all right but under high data-intensive workloads, the memory is bloated. The faults and failures mentioned above abate the profit brought by big data technology. Thus both of performance and reliability should be the important concerns when setting up a big data platform.

Performance diagnosis as the first line of defending software faults is in charge of finding out the hidden root causes of performance problems. However due to the huge cardinality of suspicious cause set, precise diagnosis in large distributed system is an extraordinarily difficult target to achieve. The difficulty is exacerbated in big data platform embodied in the following aspects.

a. Unlike the web-based applications, the execution duration of big data application is long ranging from several hours to several days (e.g. human genome analysis). Therefore the commonly used QoS metrics like response time or throughput are not suitable any more to monitor in real time. A new key performance indicator (KPI) is urgently needed.
b. The type of big data application varies a lot including both of the batch type and interactive type workloads. These two types of workloads exhibit completely different characteristics and need distinct considerations.
c. The big data platform always possesses tens of thousands of heterogeneous machines which requires the performance diagnosis approach can flexibly adapt to the scale and heterogeneity.

A wide spectrum of research has been done in this field. But most of them focus on fault location in a coarse granularity (e.g. VM or node level [3–5]). Few of them emphasize the root cause inference in a fine granularity (e.g. metric level). Recently an invariant-based performance diagnosis approach is proposed

---

[1] New properties like "Veracity" are added recently. But we still use the widely accepted three "V"s.

in [6,7] which shares a similar idea with ours. It constructs an invariant network by capturing the stable temporal and spatial relationships amongst the performance metrics collected from the whole distributed system in a pair-wise manner. This approach can work in real time and infer the root causes at fine granularity. However it's insufficient due to its workload agnostic, linear modeling and computationally intractable global invariant construction.

Considering the aforementioned challenges and the weakness of the current research, we propose a comprehensive invariant based performance diagnosis approach, *InvarNet-X*. The **goal** of *InvarNet-X* is to pinpoint the root causes for those problems whose causes are recurrent and investigated[2] and provide some hints for the unknown problems on the fly. To reduce the cost of unnecessary performance diagnosis, *InvarNet-X* first conducts the anomaly detection procedure by checking the autoregressive integrated moving average (ARIMA) model drift on CPI data of big data applications then triggers cause inference procedure. In *InvarNet-X*, each performance problem is signified by a set of violations of likely invariants constructed by MIC [10] and stored in a signature database. Finally, the real culprits are captured by searching the similar signatures in the signature database. *InvarNet-X* works under the consideration of operation context in order to adapt to the varying workloads and hardware heterogeneity. Via experimental evaluations in a small prototype, we find out *InvarNet-X* can achieve an average 91 % precision and 87 % recall in diagnosing some real faults which is superior to several state-of-the-art approaches. Our contribution is three-fold:

- We propose a new performance anomaly detection method by checking ARIMA model drift on CPI data for big data applications.
- We introduce a novel invariant construction method with MIC and build a signature database for each performance problem using the *MIC* invariant.
- We design and implement *InvarNet-X* to evaluate the accuracy and efficacy of our approach. The experimental results show that our approach can find out the culprits accurately.

The rest of this paper is organized as follows. Section 2 depicts the basic idea and problem formulation of *InvarNet-X*. Section 3 demonstrates the details of *InvarNet-X*. Section 4 shows the experimental evaluation and comparisons with several state-of-the-art approaches. Section 5 shows the related work. And Sect. 6 concludes this paper.

## 2    Problem Formulation

Our work is motivated by the methodology in medical science. The diseases have distinct behaviors from the perspective of some observable symptoms. Thus a conventional method to diagnose a disease is to look for a similar characteristic of observable symptoms from historical knowledge of investigated diseases. The historical knowledge is organized as a 'symptom-disease' database.

---

[2] These problems take up 50 %–90 % in the known performance problems [8].

In the same manner, the software system can exhibit the distinct behaviors from the viewpoint of performance metrics. The unobservable root causes of performance problems can be investigated via the directly observable runtime performance metrics. The essential work is to build the mapping function from the characteristics of performance metrics to hidden root causes. An ideal function is a one-to-one mapping. As a new exploration, Jiang [6,7] proposed an invariant-based mapping which means the performance states are characterized by the space spanned by the invariants. Our approach shares the similar idea with Jiang's work but makes some improvements. The invariants in this paper are the statistically invariant associations between performance metrics, defined as "observable likely invariant", rather than invariant statements or variables which are stated in previous study [9]. For instance, if the correlation coefficient between "used memory" and "CPU utilization" stays constant, we say these two metrics forms an invariant. Let $H$ denote the monitoring data collected from normal period and $F$ denote the monitoring data from the same system during a recent performance problem (e.g. system hang). Both of $H$ and $F$ comprise $n$ performance metrics: $(M_1, M_2, \cdots, M_n)$. We construct all the invariants of $H$ in a pair-wise manner and make them as the baseline of metric associations. These invariants are denoted by matrix $I$ where each entry $I_{M_i,M_j}$ $(i \neq j)$ represents an invariant formed by metric $M_i$ and $M_j$. Next, we use the same method to calculate the metric associations of $F$ denoted by matrix $A$, where each entry $A_{M_i,M_j}$ denotes the association between metric $M_i$ and $M_j$. If $|I_{M_i,M_j} - A_{M_i,M_j}| \geq \varepsilon$ a violation occurs where $\varepsilon$ is the preset threshold, say $\varepsilon = 0.2$ in this paper. All the violations constitute a binary tuple $(0, 1, 1, 0, \cdots, 0)$ ("0" implies no violation, "1" implies violation) which is used to signify a performance problem uniquely. The length of the tuple is determined by the number of entries in matrix $I$. Aggregating all the binary tuples constructed for multiple performance problems, a signature database is established and will be used in the future performance diagnosis. Different from Jiang's work, we adopt MIC to calculate the metric associations instead of "ARX" [6,7] due to the excellent association discovery power of MIC.

As we know, performance diagnosis is laborious and time-consuming. Hence choosing the right time to conduct performance diagnosis can reduce some unnecessary cost. In our previous work [11], we use the ARIMA model drift on several performance metrics (e.g. CPU utilization) to detect the performance anomaly. However, that method shows weak power to resist the system noise such as the resource utilization fluctuation. Therefore we set up the ARIMA model on CPI instead of other performance metrics. If a performance anomaly is detected, the cause inference procedure is triggered. We first calculate the violation tuple under the current abnormal situation then retrieve a similar signature in the signature database. If a similar signature is found, the culprit is pinpointed otherwise we provide some hints and leave the problem to the system administrators who will manually investigate the problems. Once the performance problems is resolved, a new signature will be added into the signature base.

To adapt to the varying workloads and heterogeneous hardware, we propose the concept of *"operation context"*. The operation context contains the workload type and node ID in this paper. *InvarNet-X* works under the consideration of operation context which means the performance model and signature database are built for each workload on each node.

**Restrictions:** In this paper we only validate our approach in Hadoop-based big data platforms due to its open source and widespread use. When a batch job submitted to Hadoop, Hadoop works in the FIFO mode which means the job takes up the cluster exclusively [12]. This makes *InvarNet-X* distinguish the jobs clearly. But the restriction doesn't exist when Hadoop processes interactive jobs. The performance problem is restricted on performance degradation rather than sudden crash in order to guarantee *InvarNet-X* can collect enough data to proceed diagnosis. From our previous work [2], we observe that large number of bugs can cause performance degradation such as memory leak bug. And Tan [13] also claimed 31 % bug manifested as degraded performance problem in Hadoop. Therefore our system focuses on diagnosing these problems.

Figure 1 demonstrates the basic idea of the this paper. From the figure, we can see the invariant associations between $M_1 - M_2$ and $M_2 - M_3$ on slave-3 are violated. By searching a similar signature in the signature database, we find out the root cause is a CPU-hog.

## 3    System Design

We adopt a centralized mode to implement *InvarNet-X*. Figure 3 shows the architecture of *InvarNet-X*. *InvarNet-X* leverages the performance metrics and CPI data collected from the Hadoop nodes to build the performance model, invariants

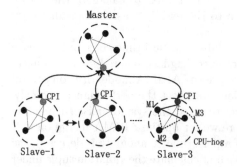

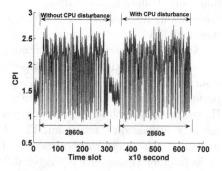

**Fig. 1.** The basic idea of *InvarNet-X*. Each small circle denotes a performance metric. The line between two performance metrics denotes the invariants and the dash line denotes the violated invariants.

**Fig. 2.** The CPI and execution time changes of *Wordcount* before and after CPU utilization disturbance. The CPU disturbance starts at 450 point and ends at 480 point

and problem signature database for each batch job and interactive job separately. The output of *InvarNet-X* is a list of root causes which puts the most probable causes in the top. The fault injection module is used to inject faults in Hadoop JobTracker, configuration files, data blocks or operating system in order to validate the effectiveness of *InvarNet-X*. In the following, we will discuss the details of *InvarNet-X*.

*InvarNet-X* mainly contains two parts and five modules shown in Fig. 3. The offline part contains three modules: performance model building, invariant construction and signature base building. The performance model building module establishes ARIMA models for specific types of workloads to describe the dynamics of CPI data. If the model on CPI data drifts, an anomaly occurs. The invariant construction module is responsible in discovering the MIC invariants amongst the performance metrics. Next, the MIC invariants are fed into the signature base building module which will find out all the violations of invariants under specific performance problems and store the violation tuples as the signatures of corresponding performance problems in a signature database. The online part contains two modules: performance anomaly detection and cause inference. When an anomaly of CPI is detected in performance detection module, the cause inference is triggered. Firstly, a violation tuple is generated by checking all the violations of invariants when the performance problem occurs. Secondly, the signatures in the signature database with a high similarity score to the violation tuple are selected. Finally the root causes corresponding to the selected signatures are reported. Compared to our previous work [11], we make several improvements on two modules including performance anomaly detection and invariant construction, other modules keep the same as before. Before we discuss the details of improvements, we first demonstrate that CPI can be a KPI of big data applications in order to detect the performance anomaly in real time.

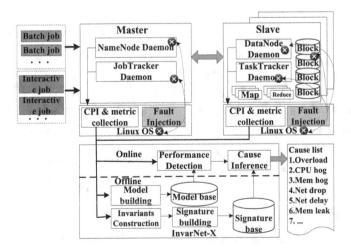

**Fig. 3.** The architecture of *InvarNet-X*

### 3.1  CPI as a KPI

In our preliminary work [11], we utilized a specific resource (e.g. CPU or memory) utilization as the KPI. It indeed indicates some performance problems in most cases. But it may mislead anomaly detection result under the disturbance of system noise. To validate this perspective, we inject resource utilization disturbance to mimic the system noise when the job (e.g. *Wordcount*) is running. From the results we observe that the execution time of some jobs have no changes although they are suffering from anomalies. Figure 2 shows the CPI changes of *Wordcount* before and after the CPU utilization disturbance (additional 30 % CPU utilization for 300 s). The CPU disturbance doesn't enlarge the execution time while the CPI keeps unaffected. Therefore a more robust KPI should be proposed to reflect the performance of the big data application.

For a specific program compiled to run on a specific machine, the execution time of this program could be expressed as:

$$T = I * CPI * C,$$

where $I$ denotes the total instructions of this program, $CPI$ denotes cycles per instruction, $C$ denotes the time length (second) of one cycle. In this equation, $I$ and $C$ are fixed. Hence the execution time $T$ only depends on $CPI$ and $CPI$ can be a candidate KPI of the big data application. To further validate this new KPI, we choose several types of jobs in BigDataBench [14] including batch type: *Wordcount, Sort, Bayes classifier* and interactive type: *TPC-DS* workloads (8 queries run in a mixed mode). 15 GB test data is generated using the BigDataBench. Four-group tests are designed. In each group, only one type of job is repeated for 25 times. And during the job running, we inject several faults such as network jam, CPU hog and disk hog to make the execution time of these jobs varies. During each time of running we collect the CPI data every 10 s and employ the

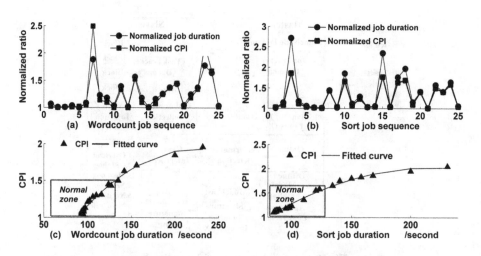

**Fig. 4.** The CPI changes with the execution time of Hadoop jobs

95 % percentile of CPI data as a sufficient statistics for one run. Other statistics like average are also applicable. For each job, the execution time and the 95 % percentile of CPI data is normalized to the minimum value respectively in one group. Due to the limited space, we only show the CPI data and execution time of *wordcount* and *sort* in Fig. 4. From Fig. 4 (a) and (b), we observe that the CPI changes with the execution time consistently. The correlation coefficient of these two metrics is 0.97 and 0.95 for *wordcount* and *sort* respectively. Figure 4 (c) and (b) demonstrate the scatter plot of CPI and execution time. And we use a 2-order polynomial function to fit these data and conclude that CPI increases monotonously with the job execution time. All the evidences show that CPI can be a stable performance indicator of big data applications. Actually, Zhang et al. [16] also utilize CPI as a performance indicator of CPU interference.

## 3.2   Performance Anomaly Detection

We employ the method proposed in our previous work [11] to detect the performance anomaly. But this paper uses CPI data rather than conventional performance metrics to build ARIMA model. If the reader wants to know the detail of ARIMA model building, please refer to [11]. The ARIMA model of CPI data in the normal state is first established and stored in an XML file in a five-tuple: *(p, d, q, ip, type)* format where the first three elements are the parameters of ARIMA, *ip* is the ip address of a Hadoop node and *type* is the workload type. To model the distinct characteristics of CPI data at "Map" and "Reduce" phases of Hadoop workloads, we utilize $N$ (e.g. 10) complete normal execution traces of CPI data of a specific type of workload to train the ARIMA model. When a new job arrives at the platform, *InvarNet-X* selects a performance model for performance anomaly detection from the archived models instantly. A simple threshold based anomaly detection method is proposed in [11]. That is if

$$\xi = |M'_{cpi}(t) - M_{cpi}(t)| > \alpha$$

a performance anomaly occurs where $M_{cpi}(t)$ is the CPI data at time $t$, $M'_{cpi}(t)$ is the CPI data predicted by ARIMA model using previous CPI data and $\alpha$ is the preset threshold. But how to set the threshold still remains a problem. In this paper, we propose three guiding rules to set the threshold. Each type of workload is repeated for $N$ times, say 20 under normal state. Next, we use the trained ARIMA model to fit the CPI data during $N$ runs. The absolute value of fitting residual is denoted by $R$. The three rules are listed below. To make the performance anomaly detection more robust to resist system noises, we report a performance problem when the anomaly occurs for three times continuously. The effectiveness of these rules will be discussed in the Sect. 4.

- **max-min**. Use $max(R)$ as the upper bar, $min(R)$ as the lower bar. If $\xi > max(R)$ or $\xi < min(R)$, an anomaly occurs.
- **95-percentile**. Use the 95 % percentile of $R$ as the threshold.
- **beta-max**. Use $\beta * max(R)$ as the threshold where $\beta$ is a fluctuation factor which is used to cover the unobserved value escaped from the test. We set $\beta = 1.2$ in this paper.

### 3.3  Invariants Construction

We use MIC to discover the association between two performance metrics. The detailed description of MIC could be found in [10]. Fore each metric pair $X, Y$, their association coefficient is represented by the $MIC(X, Y)$ score which falls in the region $[0, 1]$. In this paper, a simple but exhaustive pair-wise search method is adopted to calculate all the associations. Suppose $M$ metrics are collected from a specific node, in theory, $M(M-1)/2$ association pairs should be generated. However not all of the association pairs are invariants. The stable ones which don't fluctuate too much under the normal state are regarded as the invariants. Under the normal state, one type of workload is repeated for $N$ times. We use an association matrix to save the association pairs, denoted as $A^i$ where the superscript denotes the $i$th run and each entry $A^i(m, n)$ denotes the MIC score of metric $m$ and metric $n$. Let the vector $V(m, n) = (A^1(m, n), A^2(m, n), \cdots, A^N(m, n))$ denote the association coefficients of metric pair $(m, n)$ over $N$ runs. If a association pair doesn't exist in one run, the MIC score is assigned 0. We further select the association pairs satisfying the following condition: $Max(V(m, n)) - Min(V(m, n)) < \tau$. The threshold $\tau$ is a tunable parameter and is set 0.2 in this paper. The invariant selection algorithm is shown in Algorithm 1. When all the invariants for one type of workload are discovered, we store them in an XML file as a three-tuple $(I, ip, type)$ where $I$ stores all invariants in a matrix format, $ip$ is the ip address of a Hadoop node and $type$ is the workload type.

---

**Algorithm 1.** Invariant selection

---

**Input:** A set of performance metrics in the $N$ runs under the same workload in
  the normal state: $P^1 = (P_1^1, P_2^1, \cdots, P_M^1)$, $P^2 = (P_1^2, P_2^2, \cdots, P_M^2)$, $\cdots$, $P^N = (P_1^N, P_2^N, \cdots, P_M^N)$, $M$ is the number of performance metrics;
**Input:** A preset threshold $\tau$
**Output:** The set of invariants $I$;
1: **for** $i = 1; i = N; i + +$ **do**
2:  **for** each metric $m \in P^i$ **do**
3:   **for** each metric $n \in P^i$ **do**
4:    $A^i(m, n) = MIC(m, n)$;
5:   **end for**
6:  **end for**
7: **end for**
8: **for** each metric $m \in P^1$ **do**
9:  **for** each metric $n \in P^1$ **do**
10:   **for** $i = 1; i = N; i + +$ **do**
11:    $V(m, n) \leftarrow A^i(m, n)$ // $V(m, n)$ is a vector
12:   **end for**
13:   **if** $Max(V(m, n)) - Min(V(m, n)) < \tau$ **then**
14:    $I(m, n) \leftarrow Max(V(m, n))$
15:   **end if**
16:  **end for**
17: **end for**

---

For each performance problem whose root cause is investigated (e.g. memory leak), we build the association matrix $A_{abnormal}$ when the performance problem occurs. Then we compare $A_{abnormal}$ with the invariants $I$ of the same workload in the same Hadoop node and find out all the violations according to Sect. 2. The violations constitute a binary tuple and the tuple acts as the signature of one performance problem. The signature is stored in the signature database in the four-tuple format: *(binary tuple, problem name, ip, workload type)*. As more performance problems are diagnosed, the number of items in signature database increases gradually.

If the cause inference procedure is triggered, we adopt the approach mentioned in [11] to report the most probable root cause whose similarity score is the most close to the violation tuple. Until now the performance diagnosis is finished.

## 4    Experimental Evaluation

We have implemented a prototype and deployed it in a controlled environment. To collect the process and operating system performance metrics, a low overhead and off-the-shelf tool, *collectl*, is employed. The collected 26 performance metrics not only include coarse-grained CPU, memory, disk and network utilization but also the fine-grained metrics such as CPU context switch per second, memory page faults, etc. "perf" tool is used to collect the cycle and instruction periodically by reading the corresponding registers in the hardware performance counter on a per process basis. The collection interval is 10 s. Other parts of *InarNet-X* are developed from scratch. In the following, we will give the details of our experimental methodology and evaluation results.

### 4.1    Evaluation Methodology

Due to the lack of real operating platforms, our approach is only evaluated in a controlled big data platform. But we believe it works well in a real system without exceptions. The controlled platform contains five server machines hosting the benchmark. Each physical machine is configured with two 4-core Xeon 2.1 GHZ CPU processors, 16 GB memory, a 1 TB hard disk and a gigabit NIC and runs a 64-bit CentOS 6.2. All the servers are interconnected by a 8-port gigabit Switch. We adopt Hadoop 1.0.2, Mahout 0.6, Hive 0.9 and Mysql 5.1 as the primary software stack.

In this paper we choose four batch type of workloads: *Sort, Wordcount, Grep and Naive Bayesian classifier* and one interactive type of workloads: *TPC-DS* in BigDataBench, leaving other workloads for the future work. And the 8 queries in *TPC-DS* run simultaneously in a mixed mode. During all the experiments, 15 GB data is generated by the tool in BigDataBench benchmark. According to the reports in previous literature [13] and Hadoop bug repository [15], we inject the following faults. For the performance problems caused by runtime environment changes, we inject the following faults: (1) CPU-hog: a CPU-bound application co-locates with TaskTracker competing for CPU resource sharply; (2) Mem-hog: a memory-bound application consumes a large number of memory on one data node; (3) Disk-hog: we use a disk-bound program to generate a mass of disk reads and writes on the data node; (4) Net-drop: we use a fault injection tool "AnarchyApe" to mimic the packet loss on the name node; (5) Net-delay: we use "AnarchyApe" to delay all the packets for 800 ms; (6) Block Corruption (Block-C): we use "AnarchyApe" again to corrupt some data blocks on one data node; (7) Misconf: we set a low value (e.g. 1M) for the item "mapred.max.split.size" in the configure file; (8) Overload: we increase the current number of interactive type of workloads; (9) Suspend: we use "AnarchyApe" again to suspend the datanode or tasktracker process. For the performance problems caused by software bugs, we inject the following faults: (1) RPC-hang: the bug HADOOP-6498 causes rpc call hang. To reproduce this bug, we use hadoop inject framework to add a "sleep" statement to delay RPC call; (2) HADOOP-9703 (H-7703): when the method "stop" of "org.apache.hadoop.ipc.Client" is invoked, the thread leak happens. We use the hadoop fault inject framework to reproduce the bug by invoking this function call. (3) HADOOP-1036 (H-1036): we revert Hadoop to an older version and trigger the bug by throwing NullPointerException; (4) Lock-R: we use hadoop fault inject framework to substitute the method who has the property "synchronized" with a new method without "synchronized"; (5) HADOOP-1970 (H-1970): hadoop fault inject framework is also used to trigger this bug by interfering the communication thread; (6) Block receiver exception (Block-R): we add an exception statement in the "receivePacket" function of Class BlockReciever by hadoop inject framework. All the injected faults are guaranteed to cause significant performance problems.

Each fault mentioned above is repeated for 40 times and lasts 5 min. Two of them are used to train the signatures and the others are used to cause inference. As the probability of multiple faults happening in the same node at the same time is very tiny, we don't consider multiple faults in this paper. Actually, our method could be easily extended to multiple faults by listing multiple root causes whose signatures are most similar to the violation tuple. We leverage two commonly used metrics: precision and recall to evaluate the effectiveness of our prototype.

$$ Recall = \frac{N_{tp}}{N_{tp} + N_{fn}}, Precision = \frac{N_{tp}}{N_{tp} + N_{fp}} $$

where $N_{tp}$, $N_{fn}$, $N_{fp}$, and $N_{tn}$ denote the number of true positives, false negatives, false positives, and true negatives, respectively.

## 4.2 Performance Anomaly Detection

We use the performance anomaly detection method proposed in Sect. 3.2 to detect the anomalies incurred by fault injections. Figure 5 shows the CPI prediction residuals of *Wordcount* and *TPC-DS* using the trained ARIMA before and after CPU-hog injection. Even a cursory glance at this figure, we can see the

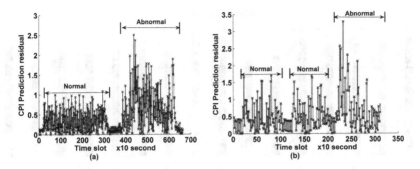

**Fig. 5.** The CPI prediction residuals before and after CPU-hog injection. (a) shows the CPI prediction residuals of workload *Wordcount*; (b) shows the CPI prediction residuals of workload *TPC-DS*

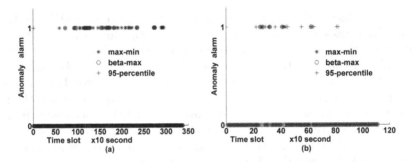

**Fig. 6.** The anomaly detection results of "max-min", "95-percentile" and "beta-max". (a) shows the results under workload *Wordcount*; (b) shows the results under workload *TPC-DS*

anomaly occurs when the CPU-hog is injected. We use the normal CPI data to train ARIMA model and use the CPI data with CPU-hog to detect anomalies. The result of anomaly detection is shown in Fig. 6 where "1" on y-axis denotes anomaly. According to the ground truth, we observe that the 95 %-percentile method has the worst detection result while the other two methods have very similar results. However the "max-min" method has a larger computational complexity than "beta-max" method due to additional "min" operation. Hence we choose "beta-max" method as the final performance anomaly detection method.

### 4.3   Diagnosis Results

We evaluate *InvarNet-X* under both of batch type of workloads and interactive type of workloads. Due to the limited space, we only show diagnosis results under workload *Wordcount* and *TPC-DS*. In reality, the diagnosis results under other workloads such as *Sort* are very similar to the shown results. Figure 7 shows the diagnosis result under workload *TPC-DS*. From this figure, we observe that

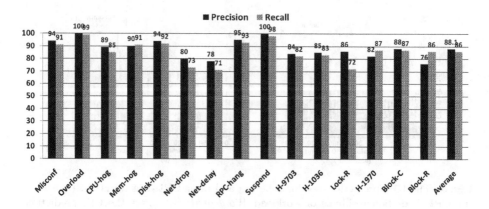

**Fig. 7.** The diagnosis result under workload *TPC-DS*

*InvarNet-X* achieves the perfect precision (100 %) and recall (99 %, 98 %) for *Overload* and *Suspend*. Because these two faults can cause a large number of violations of invariants which makes them easily distinguished from other faults by *InvarNet-X*. However the recall of *Lock-R* is very low as *Lock-R* makes different violations in different runs leading to a high false positive. For these non-deterministic problems, although *InvarNet-X* can't precisely pinpoint the root causes, it can provide some hints by showing the violated association pairs (e.g. "lock number-cpu utilization") Another interesting finding is the low accuracy of *Net-drop* and *Net-delay*. Comparing the diagnosis results with the ground truth, we find *InvarNet-X* mistakes *Net-drop* for *Net-delay* and vise versa sometimes because these two faults have very similar signatures. That's a typical "signature conflict" which will be discussed in our future work. Figure 8 shows the diagnosis result under workload *Wordcount*. When Hadoop works in FIFO mode, one job takes up the whole cluster exclusively. Therefore *overload* doesn't happen in this situation. Besides some similar characteristics with *TPC-DS*, the average precision (91.2 %) and recall (87.3 %) of *Wordcount* are higher than the average precision (88.1 %) and recall (86 %) of *TPC-DS*. That's because *TPC-DS* is a mixed workload including multiple different queries which may skew the performance model (i.e. ARIMA) and invariants even in the normal state. While *Wordcount* as a single batch job keeps a stable performance model and invariants in the normal state. In other words, the batch type of workloads possess higher quality of signatures.

Similar to our work, Jiang et al. [6,7] also propose an invariant based performance diagnosis approach. In their work, they use autoregressive models with exogenous inputs (ARX) to learn linear relationships between performance metrics. To compare with their work, we use ARX instead of MIC to implement the invariant construction. And to further validate the necessity of operation context, we implement another version of *InvarNet-X* without operation context which only contains a single performance model and signature base for one specific workload. Due to the limited space, we only show the diagnosis results

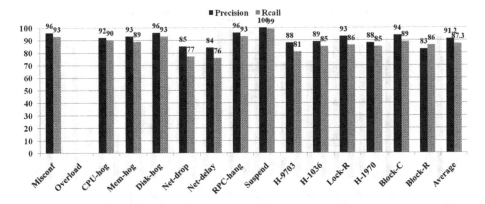

**Fig. 8.** The diagnosis result under workload *Wordcount*

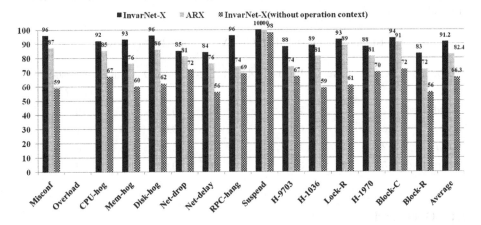

**Fig. 9.** The comparison of *InvarNet-X*, *ARX* and *InvarNet-X* (no operation context) in precision

under workload *Wordcount* in Figs. 9 and 10. Figure 9 and Fig. 10 show the diagnosis precision and recall of *InvarNet-X*, *ARX* and *InvarNet-X* (no operation context) respectively. From these two figures, we observe that the diagnosis precision of *InvarNet-X* is about 9 % higher than the one of *ARX* while the diagnosis recall shows no significant differences. The invariants discovered by *ARX* are rigorous linear relationships. The linear relationships can be broken easily when a performance problem occurs meaning that *ARX* has a strong power to capture the performance problems. However it has a weak power to distinguish the performance problems due to many similar signatures. This is the reason why we obtain the above observation. *InvarNet-X* without operation context shows a very disappointing diagnosis accuracy no matter in precision and recall. Therefore operation context is a necessary factor of performance diagnosis.

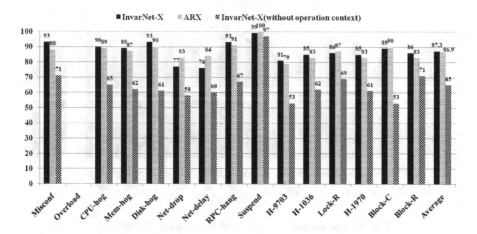

**Fig. 10.** The comparison of *InvarNet-X*, *ARX* and *InvarNet-X* (no operation context) in recall

**Table 1.** The overhead of *InvarNet-X* and *ARX* (/second)

| Workload | Perf-M | Invar-C | Invar-C (*ARX*) | Sig-B | Perf-D | Cause-I | Cause-I (*ARX*) |
|---|---|---|---|---|---|---|---|
| Wordcount | 1.2 | 45 | 700 | 4 | 0.02 | 1.6 | 10 |
| Sort | 0.8 | 30 | 650 | 3 | 0.03 | 1.7 | 11 |
| Grep | 0.2 | 18 | 410 | 1.7 | 0.02 | 1.6 | 10 |
| Interactive | 0.5 | 16 | 380 | 1.5 | 0.03 | 1.6 | 12 |

### 4.4   Overhead

Here we only consider the CPU overhead because other types of overhead like memory and disk caused by *InvarNet-X* are very small. The CPU overhead contains six parts: data collection, performance model building (Perf-M), invariant construction (Invar-C), signature building (Sig-B), performance anomaly detection (Perf-D) and cause inference (Cause-I). The data collection costs no more than 5 % CPU utilization. Table 1 shows the execution time of the other five parts under different types of workloads. We observe that the execution time of Perf-D and Cause-I stays below 2 s satisfying the online requirement. While the execution time of Cause-I(*ARX*) is around 10 s much larger than Cause-I(*InvarNet-X*). Although as an offline part, the execution time Invar-C (*InvarNet-X*) is up to 45 s, it is much lower than the one of Invar-C (*ARX*) in one order of magnitude. Therefore *InvarNet-X* is computationally tractable when it scales up in large scale big data platform.

## 5    Related Work

A large quantity of work has been done in performance diagnosis of general distributed systems. However most of them are concerned with fault location in a coarse granularity (e.g. VM level [3–5]). Few of them emphasize the root cause inference in a fine granularity. Towards performance diagnosis in MapReduce (e.g. hadoop) system, the work could be roughly categorized into two classes: log-based and correlation-based. The log-based method [13] can pinpoint the buggy code in a very fine granularity (e.g. line of code) but it is hard to conduct in real time. The correlation-based method [5] more often than not uses the *peer-similarity* to find out the abnormal nodes assuming that the correlations amongst the performance metrics of different nodes are stable. However an exceptional case exists. Assume one bug exists in the platform, when the bug is triggered by a certain job, say *wordcount*, all the nodes behave abnormally in a similar way but the correlations are not deviated. In this case, the correlation-based method will ignore this fault.

Recently an invariant-based performance diagnosis approach is proposed in [6,7]. It constructs an invariant network by capturing the stable temporal and spatial relationships amongst the performance metrics in a pair-wise manner. A set of deviations of these invariants can indicate a specific fault. This approach can work in real time and infer the root causes at fine granularity. However this method has the following limitations: (a) It is workload agnostic. As pointed in [7], they selected 111 measurements form the system and 74 of them are correlated with the workload. Moreover according to our work [11], it's hard to find out such a model suitable to all kinds of workloads. Hence workload is an important factor in performance diagnosis. (b) It only considers linear relationships between performance metrics leading to invariant missing. Due to highly dynamical nature of software system, non-linearity is a more common case. (c) The global constructions of invariant network and simultaneously checking all the invariants in real time make it computationally intractable in the large scale distributed environment.

## 6    Conclusion

This paper proposes a comprehensive invariant-based approach, *InvarNet-X*, to pinpoint the culprits of performance problems in the big data platform. *InvarNet-X* not only covers performance anomaly detection but also root cause inference. The performance anomaly procedure is accomplished by checking the ARIMA model drift on CPI data of big data applications. In *InvarNet-X*, the likely invariants are established via MIC and each performance problem is signified by a set of violations of those likely invariants. Finally, the root cause is uncovered by searching a similar signature in the signature database. Through experimental evaluations in a small prototype, we find out *InvarNet-X* can achieve an average 91 % precision and 87 % recall in diagnosing some real faults which is superior to several state-of-the-art approaches. Meanwhile *InvarNet-X* causes a low overhead to the system.

**Acknowledgments.** We thank to all the members in our research group.

# References

1. Dean, J., Ghemawat, S.: MapReduce: simplified data processing on large clusters. Commun. ACM **51**(1), 107–113 (2008)
2. Chen, P., Qi, Y., Hou, D., Zheng, P.: CauseInfer: automatic and distributed performance diagnosis with hierarchical causality graph in large distributed systems. In: 33rd Annual IEEE International Conference on Computer Communications, Toronto (2014)
3. Bodik, P., Goldszmidt, M., Fox, A., Woodard, D.B., Andersen, H.: Fingerprinting the datacenter: automated classification of performance crises. In: 5th European Conference on Computer Systems, pp. 111–124. ACM Press, Lancaster (2010)
4. Nguyen, H., Shen, Z., Tan, Y., Gu, X.: FChain: toward black-box online fault localization for cloud systems. In: 33rd International Conference on Distributed Computing Systems (ICDCS), pp. 21–30. IEEE Press, Philadelphia (2013)
5. Kang, H., Chen, H., Jiang, G.: PeerWatch: a fault detection and diagnosis tool for virtualized consolidation systems. In: 7th International Conference on Autonomic Computing, pp. 119–128. ACM Press, London (2010)
6. Jiang, G., Chen, H., Yoshihira, K.: Efficient and scalable algorithms for inferring likely invariants in distributed systems. IEEE Trans. Knowl. Data Eng. **19**(11), 1508–1523 (2007)
7. Jiang, G., Chen, H., Yoshihira, K.: Discovering likely invariants of distributed transaction systems for autonomic system management. In: 3rd IEEE International Conference on Autonomic Computing, pp. 199–208. ACM Press, New York (2006)
8. Duan, S., Babu, S., Munagala, K.: Fa: a system for automating failure diagnosis. In: 25th IEEE International Conference on Data Engineering, pp. 1012–1023. IEEE Press, Shanghai (2009)
9. Ernst, M.D., Perkins, J.H., Guo, P.J., McCamant, S., Pacheco, C., Tschantz, M.S., Xiao, C.: The Daikon system for dynamic detection of likely invariants. Sci. Comput. Program. **69**(1), 35–45 (2007)
10. Reshef, D.N., Reshef, Y.A., Finucane, H.K., Grossman, S.R., McVean, G., Turnbaugh, P.J., Sabeti, P.C.: Detecting novel associations in large data sets. Science **334**(6062), 1518–1524 (2011)
11. Chen, P., Qi, Y., Li, X., Su, L.: An ensemble MIC-based approach for performance diagnosis in big data platform. In: 1st IEEE International Conference on Big Data, pp. 78–85. IEEE Press, Santa Clara (2013)
12. Sangroya, A., Serrano, D., Bouchenak, S.: Benchmarking dependability of MapReduce systems. In: 31st IEEE International Symposium on Reliable Distributed Systems, pp. 21–30. IEEE Press, Irvine (2012)
13. Tan, J., Pan, X., Marinelli, E., Kavulya, S., Gandhi, R., Narasimhan, P.: Kahuna: problem diagnosis for MapReduce-based cloud computing environments. In: 12th IEEE/IFIP Network Operations and Management Symposium, pp. 112–119. IEEE Press, Osaka (2010)
14. Wang, L., Zhan, J., Luo, C., et al.: BigDataBench: a big data benchmark suite from internet services (2014). arXiv preprint arXiv:1401.1406
15. Hadoop bug repository. http://hadoop.apache.org/issue_tracking.html
16. Zhang, X., Tune, E., Hagmann, R., et al.: CPI2: CPU performance isolation for shared compute clusters. In: 8th ACM European Conference on Computer Systems, pp. 379–391. ACM Press, New York (2013)

# Tuning Hadoop Map Slot Value Using CPU Metric

Kamal Kc[⊠] and Vincent W. Freeh

North Carolina State University, Raleigh, USA
{kkc,vwfreeh}@ncsu.edu

**Abstract.** Hadoop is a widely used open source mapreduce framework. Its performance is critical because it increases the usefulness of products and services for a large number of companies who have adopted Hadoop for their business purposes. One of the configuration parameters that influences the resource allocation and thus the performance of a Hadoop application is map slot value (MSV). MSV determines the number of map tasks that run concurrently on a node. For a given architecture, a Hadoop application has an MSV for which its performance is best. Furthermore, there is not a single map slot value that is best for all applications. A Hadoop application's performance suffers when MSV is not the best. Therefore, knowing the best MSV is important for an application. In this work, we find a low-overhead method to predict the best MSV using a new Hadoop counter that measures per-map task CPU utilization. Our experiments on a variety of Hadoop applications show that using a single MSV for all applications results in performance degradation up to 132 % when compared to using the best MSV for each application.

## 1 Introduction

Hadoop is an open source mapreduce framework used by hundreds of companies for a variety of applications, which include indexing products in ecommerce webservices, log analysis, reporting, analytics, and machine learning [2]. The performance of Hadoop is important in increasing the usefulness of these products.

Performance tuning in Hadoop is a complex task as it has more than 150 configuration parameters that directly or indirectly affect its resource utilization and performance. The most common method for selecting best configuration values is trying several possible values and manually tweaking them until a Hadoop application completes in the least amount of time [1]. This process quickly becomes cumbersome and inefficient when finding best values for more than one Hadoop application. Thus, it is desirable to have a mechanism to select a best set of configuration parameters. In this work, we find a mechanism to predict the best value of a Hadoop parameter called *map slot value*.

Map slot value (MSV) is the maximum number of map tasks that run concurrently on a tasktracker node. Its misconfiguration can create significant performance degradation. We find that a Hadoop application has a select best MSV or MSVs for which its performance is the best. Furthermore, there is not a single

© Springer International Publishing Switzerland 2014
J. Zhan et al. (Eds.): BPOE 2014, LNCS 8807, pp. 141–153, 2014.
DOI: 10.1007/978-3-319-13021-7_11

MSV that has the best performance for all Hadoop applications. For example, in one of our experimental clusters there are four MSVs that have the best performance for at least one application, but these MSVs have maximum performance degradation as high as 132 %. Thus, it is important to know the best MSV for an application in order to avoid its performance degradation.

In this paper, we present a method to reliably predict the best MSV with low-overhead. Our method uses two new Hadoop counters that measure per-map task CPU utilization and IO throughput. In the following sections of the paper, we present the related work, describe Hadoop map phases, describe our modifications, present the map phase completion time results, and describe the prediction of MSV.

## 2   Related Work

Four areas of prior research are related to our work. The first area concerns optimizing configuration parameters of Hadoop. Research in this area explores different methods to obtain the optimal Hadoop configuration values. The methods include deriving the values from the optimal values of other jobs [9] or using metrics obtained from extensive instrumentation to extrapolate the optimal values [5,7,8]. Our work does not require knowing optimal values of other jobs or performing extensive instrumentation.

The second area of prior research is the optimization of resource utilization by selecting the best predefined technique. These approaches offer an important alternative method for optimizing performance behavior. They select the best technique by using rules [3], using program analysis [11] or measuring the completion time of several alternative implementations [13].

The third area of prior research is workload analysis of Hadoop jobs. These studies focus on creating standardized benchmarks [6] or creating qualitative job classes such as small, medium, and large duration jobs [10]. Our work can be extended to correlate the previously studied job classes and their performance behavior.

The fourth area that is closely related to our work is the research on Hadoop schedulers. This work does not focus on optimizing the resource utilization of the entire cluster, but rather focuses on optimizing resource utilization within the resource bounds imposed by the fixed preconfigured MSV [12,14]. Our approach finds the best MSV, which ensures the efficient utilization of the cluster resources.

In addition to prior research, PUMA is also used in our work [4]. PUMA is a Hadoop benchmarking suite developed by Purdue University. PUMA includes three mapreduce programs from the official Hadoop distribution and ten other mapreduce programs. The collection of diverse applications makes it a useful benchmarking suite. In our experiments, we use six PUMA Hadoop applications. The remainder of the PUMA applications have similar characteristics on the metrics we measure. The combination of the six with our custom applications include the entire range of the measured metrics. In our future work, we plan to evaluate additional benchmark programs, including all those in PUMA.

# 3  Hadoop and Modifications

A Hadoop application consists of *map* and *reduce* tasks. A map task processes a block of data and produces key-value output pairs. The map output is partitioned according to the range of the key. A reduce task aggregates and operates on the map output key-value pairs that fall under the assigned key range partition.

The map phase is usually the most time consuming operation of a Hadoop application. The applications used in our work, which represent common Hadoop applications, have an average map time of 67 % of the total job runtime. The lowest map time for an application is 42 % of the total job runtime whereas the highest map time is 85 % of the total job runtime. Due to this reason, in this work we focus on optimizing the map runtime.

## 3.1  Map Phases

A map task consists of 6 phases, which are compute, collect, sort, spill, combine, and merge-spill. In the compute phase, a map task applies the map function to each input key-value pair. In the collect phase, the map task stores the processed key-value pairs in a map output buffer. The sort phase occurs between collect and spill operations. In this phase, the output key-value pairs are sorted before the spilling occurs. When the map output buffer is full, the map task empties the buffer by spilling its content to a spill file in the local disk. This is the spill phase. The combine phase is optional and when present, the map task performs a local reduce operation on the map output key-value pairs. In the merge-spill phase, the spill files are merged together to produce a single map output file.

Each phase in the map task is either CPU or IO intensive. The CPU intensive phases are compute, collect, sort, and combine. The IO intensive phases are spill and merge-spill.

## 3.2  Modifications

Our approach predicts the best MSV using the CPU utilization and IO through-put metrics of a map task. The metrics are derived from the durations of the map phases. The best MSV is the MSV setting for which an application has the short-est completion time. An application has the best performance when all CPUs are fully utilized. Additionally, IO bound applications suffer performance degrada-tion when the IO bandwidth is fully consumed. Thus, the best MSV setting either utilizes all CPUs efficiently or in case of an IO bound application fully utilizes the IO bandwidth. For MSVs greater than the best, the parallelism is too great, resulting in either CPU or IO performance degradation. The CPU performance degradation occurs due to additional system overhead for the larger number of processes. The IO performance degradation occurs due to higher contention for the IO bandwidth which lowers the overall IO throughput of the system. On the other hand, for MSVs lower than the best, the parallelism is low, resulting in either CPU or IO underutilization. When the CPU is underutilized, the CPU user and system time is low. When the IO resource is underutilized, there is leftover IO bandwidth available for use. In both cases, additional tasks can be run to use the leftover resources and improve the application completion time.

We use Hadoop counters to measure per-map task CPU utilization and IO throughput. Counters are built-in low-overhead metrics in Hadoop. They report important task and job related statistics. By default, a Hadoop task collects at least 16 statistical values using the counters. An example of a Hadoop counter is HDFS_BYTES_WRITTEN, which records the total amount of output data written to HDFS by reduce tasks.

We introduce two new counters CPU_UTIL and IO_THRPUT to measure per-map task CPU utilization and IO throughput. CPU_UTIL is the sum of the time taken by the CPU intensive phases of a map task, which are compute, collect, sort, and combine. IO_THRPUT is the quotient of total map output bytes divided by the map task duration. The overhead of these counters is insignificant and is same as maintaining other existing Hadoop counters. Additionally, as map tasks have homogeneous CPU and IO behavior, we only need to measure the counter values of a map task instance to estimate the best MSV for an application.

## 4  Evaluation

In this section, we describe the experimental setup, analyze the performance of the Hadoop applications for different MSVs, and describe the prediction of MSV using the metric values.

### 4.1  Experimental Setup

Experiments are performed on two clusters. The first cluster consists of six IBM PowerPC machines. Each node contains two POWER7 processors with 24 cores and 48 total CPU threads, 90 GB RAM, and a 10 Gbps Ethernet network link. In the PowerPC cluster, Hadoop is configured with one jobtracker and five task-trackers. HDFS is configured with one namenode and five datanodes. The second cluster consists of six x86 machines. Each node contains two Intel Xeon x86 processors with 8 cores and 16 total CPU threads, 24 GB RAM, and a 10 Gbps Ethernet link. As in the PowerPC, in this cluster, Hadoop is configured with one jobtracker and five tasktrackers. HDFS is configured with one namenode and five datanodes.

Our experiments use thirteen Hadoop applications, among which six applications are from PUMA benchmark suite [4] and the remaining seven applications are customized versions of *terasort*. The PUMA applications are *grep*, *wordcount*, *invertedindex*, *rankedinvertedindex*, *terasort*, and *termvectorperhost*. The PUMA applications use wikipedia dataset. Terasort and its variants use the data generated by teragen.

We use the variants of terasort in order to explore the entire spectrum of CPU utilization and IO throughput values. Among the PUMA applications, the lowest CPU utilization of a map task is 36 % for terasort and the highest is

**Table 1.** Utilization and throughput in the ascending order of CPU_UTIL for PowerPC cluster.

| Applications | CPU_UTIL (%) | IO_THRPUT (MB/s) |
|---|---|---|
| terasort | 36 | 7.31 |
| rankedinvertedindex | 40 | 4.42 |
| terasort(L10, D100) | 46 | 4.79 |
| terasort(L30, D100) | 58 | 3.87 |
| wordcount | 58 | 2.84 |
| invertedindex | 65 | 1.99 |
| terasort(L60, D100) | 69 | 3.02 |
| termvectorperhost | 75 | 3.77 |
| terasort(L100, D100) | 77 | 2.29 |
| terasort(L200, D100) | 87 | 1.41 |
| terasort(L500, D100) | 94 | 0.65 |
| terasort(L10, D1) | 97 | 0.11 |
| grep | 97 | 0.01 |

97 % for grep. Similarly, the lowest IO throughput is 0.01 MB/s for grep and the highest is 7.31 MB/s for terasort. However, the PUMA applications do not include all CPU utilization values between 36 % and 97 % or all IO throughput values between 0.01 MB/s and 7.31 MB/s. In order to include the entire utilization spectrum, in the terasort application we add a variable number of extra busy loops and to include the entire IO throughput spectrum we vary the amount of map output data. Table 1 shows that after adding the terasort variants, the applications include the entire spectrum of CPU utilization from 36 % to 99 % and IO throughput from 0.01 MB/s to 7.31 MB/s. The terasort variants are listed by showing the number of busy loops and the amount of output data. For the terasort variant terasort(L10, D100), L represents the number of busy loops and D represents the percentage of input data that is converted to output. Thus, terasort(L10, D100) executes 10 extra busy loops for each key-value pair and outputs 100 % of the input data. The highest number of busy loops is 500 and the lowest amount of output data is 0.01 %. The variables L and D are used to control the CPU utilization and IO throughput of a map task. Increasing the value of L increases the amount of CPU computation performed by a map task. Additionally, decreasing the value of D decreases the amount of output data produced by a map task.

## 4.2   Performance Analysis

Table 2 shows the normalized performance behavior of the thirteen applications for different MSVs and normalized MSVs for PowerPC cluster with 150 GB data. The normalized MSV is the MSV relative to the number of CPU threads in a

**Table 2.** Normalized performance and the best completion time (in parentheses) for different MSVs on the PowerPC cluster with 150 GB datasize.

| Data size = 150 GB, CPU cores per node = 24, CPU threads per node = 48 | | | | | | | |
|---|---|---|---|---|---|---|---|
| Job | MSV(normalized) | | | | | | |
| | 16 (0.33) | 24 (0.50) | 32 (0.67) | 40 (0.83) | 48 (1) | 56 (1.17) | 64 (1.33) |
| terasort | 1.17 | **1(258 s)** | 1.02 | 1.18 | 1.89 | 2.32 | 2.51 |
| rankedinvertedindex | 1.30 | 1.09 | 1.01 | **1(453 s)** | 1.11 | 1.09 | 1.23 |
| terasort(L10, D100) | 1.28 | 1.07 | 1.01 | **1(350 s)** | 1.16 | 1.03 | 1.19 |
| terasort(L30, D100) | 1.34 | 1.20 | 1.14 | 1.08 | 1.19 | **1(470 s)** | 1.17 |
| word count | 1.57 | 1.26 | 1.26 | 1.06 | 1.08 | **1(564 s)** | 1.04 |
| invertedindex | 1.49 | 1.21 | 1.21 | 1.06 | 1.09 | **1(620 s)** | 1.02 |
| terasort(L60, D100) | 1.19 | 1.13 | 1.15 | 1.11 | 1.14 | **1(689 s)** | 1.13 |
| termvectorperhost | 1.38 | 1.16 | 1.16 | 1.02 | 1.08 | **1(694 s)** | 1.02 |
| terasort(L100, D100) | 1.15 | 1.08 | 1.10 | 1.07 | 1.01 | **1(948 s)** | 1.09 |
| terasort(L200, D100) | 1.16 | 1.10 | 1.10 | 1.10 | 1.13 | **1(1539 s)** | 1.12 |
| terasort(L500, D100) | 1.15 | 1.08 | 1.11 | 1.09 | 1.06 | **1(3459 s)** | 1.05 |
| terasort(L10, D1) | 1.13 | **1(173 s)** | 1.01 | 1.20 | 1.59 | 1.90 | 1.98 |
| grep | 1.18 | 1.05 | 1.05 | **1(245 s)** | 1.11 | 1.39 | 1.40 |
| **Average** | **1.27** | **1.11** | **1.11** | **1.07** | **1.20** | **1.21** | **1.30** |
| **# of best values** | 0 | 2 | 0 | 3 | 0 | 8 | 0 |

node.[1] The normalized MSVs are shown in parentheses alongside the MSVs in the header of the table. For the PowerPC machines, a normalized MSV of 1 means an actual MSV of 48, which is equal to 1 map task per CPU thread (or 2 map tasks per core). The best performance value is 1 and it denotes the shortest completion time of an application for the set of MSVs used in the experiments. The best MSV is the one for which an application has the shortest completion time. The shortest completion time is shown in parentheses for each application. In the PowerPC cluster, MSV is set to values from 16 to 64 in increments of 8. Below 16 and beyond 64, the applications suffer slowdown and those results are omitted.

Table 2 shows that there is a best value for each application and there is not a single MSV that is best for all applications. Every application in the table has select best MSV or MSVs (six applications have MSV ≤ 1.02). For example, terasort and wordcount have best MSVs of 24 and 56. Additionally, the table

---

[1] The machines used in our experiments have hyperthreading enabled. Thus the CPU schedulable contexts as seen by operating system is greater than the number of cores. In a hyperthreaded system, the metric CPU_UTIL measures the utilization of the threads, and a 100% utilization occurs when all threads are busy rather than when all cores are busy.

**Table 3.** Normalized performance the best completion time (in parentheses) for different MSVs on PowerPC cluster with 300 GB datasize.

| Data size = 300 GB, CPU cores per node = 24, CPU threads per node = 48 | | | | | | | |
|---|---|---|---|---|---|---|---|
| Job | MSV(normalized) | | | | | | |
| | 16 | 24 | 32 | 40 | 48 | 56 | 64 |
| | (0.33) | (0.50) | (0.67) | (0.83) | (1) | (1.17) | (1.33) |
| terasort | 1.13 | **1(519 s)** | 1.06 | 1.57 | 2.42 | 2.49 | 2.92 |
| rankedinvertedindex | 1.32 | 1.09 | 1.02 | **1(771 s)** | 1.22 | 1.47 | 2.09 |
| terasort(L10, D100) | 1.27 | 1.09 | **1(663 s)** | 1.03 | 1.42 | 1.14 | 1.73 |
| terasort(L30, D100) | 1.32 | 1.12 | 1.10 | 1.04 | 1.02 | **1(939 s)** | 1.11 |
| word count | 1.52 | 1.22 | 1.09 | 1.03 | 1.03 | **1(1110 s)** | 1.02 |
| invertedindex | 1.49 | 1.20 | 1.09 | 1.03 | 1.02 | 1.01 | **1(1307 s)** |
| terasort(L60, D100) | 1.19 | 1.14 | 1.09 | 1.05 | 1.03 | **1(1328 s)** | 1.09 |
| termvectorperhost | 1.35 | 1.18 | 1.12 | 1.07 | 1.02 | **1(1368 s)** | 1.03 |
| terasort(L100, D100) | 1.17 | 1.15 | 1.11 | 1.04 | 1.03 | **1(1800 s)** | 1.09 |
| terasort(L200, D100) | 1.16 | 1.14 | 1.13 | 1.07 | 1.05 | **1(3030 s)** | 1.06 |
| terasort(L500, D100) | 1.14 | 1.13 | 1.11 | 1.09 | 1.05 | **1(6814 s)** | 1.07 |
| terasort(L10, D1) | 1.08 | **1(248 s)** | 1.01 | 1.02 | 1.20 | 1.17 | 1.23 |
| grep | 1.33 | 1.17 | 1.05 | **1(513 s)** | 1.01 | 1.06 | 1.03 |
| **Average** | **1.27** | **1.13** | **1.07** | **1.08** | **1.19** | **1.18** | **1.34** |
| **# of best values** | **0** | **2** | **1** | **2** | **0** | **7** | **1** |

does not have a single MSV that is best for all applications. The last row shows the number of best values for different MSVs. MSVs 24, 40, and 56 are best for 2, 3, and 8 applications. One significant MSV is 40, which has lowest average performance value of 1.07. This is 7 % higher than the theoretical best performance value of 1. But, it has a maximum slowdown of 20 % for terasort(L10, D10). Thus, picking MSV 40 for all applications is not an adequate best solution. Additionally, while MSV of 56 is best for the most applications, it has a slowdown of 132 % for terasort. This further reinforces that a single MSV is not a best choice for all applications.

In order to test if these results are generally applicable, we also run these applications on both a larger dataset size and a different architecture (x86). Table 3 shows the normalized performance behavior of the PowerPC cluster for 300 GB dataset. The results show performance behavior similar to the PowerPC cluster with 150 GB data size. In Table 3, MSVs 24, 32, 40, and 56 are best for at least one application. MSV 32 has the lowest average performance value of 1.07. It has a maximum slowdown of 13 %.

Table 4 shows the normalized performance behavior of the x86 cluster for 150 GB dataset. It shows that MSVs 8, 12, and 16 are best for at least one application. As the number of cores and threads are different in the x86 cluster, for the experiment, MSV is set to values from 4 to 24 with increments of 4. Beyond 24, the applications suffer slowdown. MSV 12 has the lowest average

**Table 4.** Normalized performance the best completion time (in parentheses) for different MSVs on x86 cluster with 150 GB datasize.

| Data size = 150 GB, CPU cores per node = 8, CPU threads per node = 16 | | | | | | |
|---|---|---|---|---|---|---|
| Job | MSV(normalized) | | | | | |
| | 4 | 8 | 12 | 16 | 20 | 24 |
| | (0.25) | (0.50) | (0.75) | (1) | (1.25) | (1.5) |
| terasort | 1.14 | **1(2688 s)** | 1.05 | 1.11 | 1.34 | 1.46 |
| rankedinvertedindex | 1.09 | **1(1676 s)** | 1.11 | 1.24 | 1.35 | 1.44 |
| terasort(L10, D100) | 1.13 | **1(2172 s)** | 1.09 | 1.21 | 1.31 | 1.44 |
| terasort(L30, D100) | 1.39 | 1.12 | **1(2670 s)** | 1.05 | 1.11 | 1.42 |
| word count | 1.88 | 1.04 | **1(1337 s)** | 1.17 | 1.12 | 1.21 |
| invertedindex | 2.36 | **1(509 s)** | 1.09 | 1.41 | 1.95 | 2.13 |
| terasort(L60, D100) | 1.24 | 1.24 | 1.01 | **1(2732 s)** | 1.17 | 1.41 |
| termvectorperhost | 1.56 | 1.10 | 1.09 | **1(536 s)** | 1.13 | 1.20 |
| terasort(L100, D100) | 1.61 | 1.29 | 1.14 | **1(2871 s)** | 1.23 | 1.40 |
| terasort(L200, D100) | 1.54 | 1.18 | 1.13 | **1(3114 s)** | 1.11 | 1.21 |
| terasort(L500, D100) | 1.85 | 1.25 | 1.16 | **1(3960 s)** | 1.09 | 1.27 |
| terasort(L10, D1) | 1.04 | **1(883 s)** | 1.05 | 1.08 | 1.06 | 1.07 |
| grep | 1.57 | 1.12 | **1(352 s)** | 1.05 | 1.09 | 1.10 |
| **Average** | **1.49** | **1.1** | **1.08** | **1.11** | **1.24** | **1.37** |
| **# of best values** | 0 | 5 | 3 | 5 | 0 | 0 |

performance value of 1.08, which is 8 % higher than the best average performance value. It has a maximum slowdown of 16 %. Thus, in the three cases tested, there is not a single MSV that is best for all applications.

### 4.3   Prediction

Table 1 shows an inverse linear relationship between CPU_UTIL and IO_THR-PUT metrics. When CPU_UTIL increases, IO_THRPUT decreases and *vice versa*. Using linear regression of Table 1 data, we can approximate IO_THRPUT for the PowerPC cluster. Therefore, while searching for a metric to predict best MSV for a Hadoop application, we only use the values of CPU_UTIL. Figure 1 shows the normalized best MSVs for the CPU utilization combinations of all 13 applications.

In Fig. 1, the applications are divided into three general regions: *IO-intensive*, *Balanced*, and *CPU-intensive*. Each region is based on a different range of CPU_UTIL and a predictable range of best MSV. *IO-intensive* region has applications with low CPU_UTIL (36 %–60 %). This region is called *IO-intensive* because applications in this region have high IO_UTIL. The normalized best MSVs of applications in this region are less than 1.0. *Balanced* region has

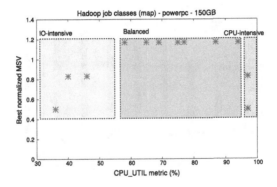

**Fig. 1.** Classification for applications running on PowerPC cluster.

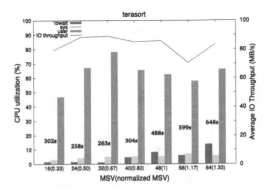

**Fig. 2.** CPU utilization, IO throughput, and completion time (shown in relative vertical heights) of an *IO-intensive* application (terasort).

applications with medium CPU_UTIL (60 %–90 %). The normalized best MSVs of applications in this region are greater than 1.0. *CPU-intensive* region has applications with high CPU_UTIL (90 %–100 %). Applications in this region have low IO_UTIL. The best normalized MSVs of applications in this region are less than 1.0. The normalized best MSVs greater than 1.0 means that the number of map tasks exceeds the number of CPU threads in the system. This indicates that the applications are scalable. On the other hand, the best normalized MSVs less than 1.0 indicate that the applications are not scalable and face resource bottlenecks. The scalable nature of *Balanced* region and the bottlenecks of *IO-intensive* and *CPU-intensive* regions are described with examples in the following paragraphs.

Figures 2, 3, and 4 show the CPU and IO behavior of a tasktracker node for an application of each region. The figures show the node behavior including the completion time for all MSVs and help to explain the performance of an application when MSV is the best. In the figures, the CPU utilization is divided into *user*, *system*, and *iowait* states. The metric *user* is the time spent by the map

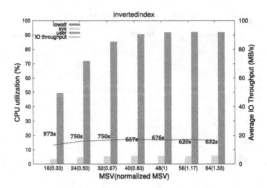

**Fig. 3.** CPU utilization, IO throughput, and completion time (shown in relative vertical heights) of a *Balanced* application (invertedindex).

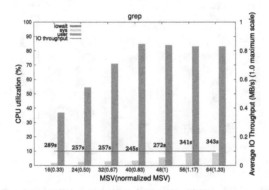

**Fig. 4.** CPU utilization, IO throughput, and completion time (shown in relative vertical heights) of a *CPU-intensive* application (grep).

tasks, *system* is the time spent by kernel, and *iowait* is the time spent waiting for IO operation to complete. Each figure is described as follows.

Figure 2 shows the node behavior of *terasort*, which falls in *IO-intensive* region. The figure shows 24 as the best MSV. In Fig. 2, the IO throughput increases until MSV is 24. For MSV greater than 24, the IO throughput decreases and there is an increase in *iowait*. This suggests that beyond 24 MSV, increasing parallelism merely increases IO pressure and overhead due to the IO pressure. This explains the lower relative best MSV for applications in *IO-intensive* region.

Figure 3 shows the node behavior of *invertedindex*, which falls in *Balanced* region. The figure shows 56 as the best MSV. In Fig. 3, the *user* CPU increases until MSV is 56. After 56, the *user* CPU levels off without showing performance improvement. IO throughput on the other hand is almost constant and does not peak, which suggests a lack of IO bottleneck. Due to this reason, there is not any noticeable *iowait*. This behavior results in the applications having best MSV greater than the number of CPU threads.

Figure 4 shows the node behavior of *grep*, which falls in *CPU-intensive* region. The figure shows 40 as the best MSV, which is 83 % of the total number of virtual CPU threads. In Fig. 4, the *user* CPU steadily increases until MSV is 40. After 40, the *sys* CPU increases and the *user* CPU levels off and decreases in small amount. This suggests that due to the high CPU utilization of grep, the system overhead increases. The IO throughput on the other hand is 0.1 MB/s for all MSVs.

**Prediction for x86 cluster.** The prediction for x86 cluster is similar to the PowerPC, except the difference in region boundaries. The IO throughput of a x86 node and a PowerPC node is 40 MB/s and 100 MB/s respectively.[2] Due to this reason, applications with medium CPU_UTIL, which in turn have medium IO_THRPUT, suffer from IO bottleneck in the x86 cluster whereas they do not suffer from IO bottleneck in the PowerPC cluster. As a result, the applications with medium CPU_UTIL belong to *IO-intensive* region of x86 cluster instead of *Balanced* region. Additionally, as these applications with medium IO_THRPUT have relatively higher CPU_UTIL values, the *Balanced* region starts at a higher CPU_UTIL value in the x86 cluster. This is observed in Fig. 5, which shows the best normalized MSV for all CPU_UTIL values. From the figure, we observe that the *Balanced* region starts at 69 % which is relatively higher than 58 % for PowerPC.

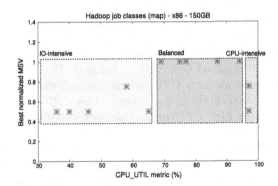

**Fig. 5.** Classification for applications running on x86 cluster.

**Performance characteristics using predicted MSV.** To find the effectiveness of the regions, we compare the performance of applications when using the region specific MSVs and a best single MSV for all applications. For the *IO-intensive*, *Balanced*, and *CPU-intensive* regions, we select the normalized MSVs 0.67, 1.17, and 0.67 for the PowerPC cluster and 0.75, 1, and 0.75 for the x86 cluster. The best single MSV for PowerPC is 0.67 and for x86 it is 0.75. Table 5

---

[2] A node in the x86 cluster has a single SAS hard disk, whereas a PowerPC node has 5 SAS hard disks in RAID-5 configuration.

**Table 5.** Average normalized performance values and maximum slowdown percentages (inside parentheses) for the three tested cases.

| MSV selection | PowerPC | | x86 |
|---|---|---|---|
| | 150 GB | 300 GB | 150 GB |
| Best single MSV | 1.11(20 %) | 1.07(13 %) | 1.08(16 %) |
| Predicted MSV using the regions | 1.01(6 %) | 1.01(6 %) | 1.03(11 %) |

shows the performance values for these two schemes. Using a single predicted MSV for applications in each region has a better aggregate performance value. For the three cases, the aggregate slowdown when using region based predicted MSV compared to when using a single fixed MSV is 1 % and 11 %, 1 % and 7 %, and 3 % and 8 % respectively. In the parentheses alongside the performance values, the table shows in percentage the maximum slowdown of applications when using the predicted MSVs. It gives an upper bound of slowdown for an application, which occurs during the worst case scenario. By using the region based predicted MSVs, the maximum slowdown decreases to as little as 6 % from 20 % when compared to using a single best MSV.

In this section, we showed the performance results of Hadoop applications for two clusters: PowerPC and x86, and two data sizes for PowerPC cluster: 150 GB and 300 GB. Our findings show that we can predict the performance behavior based on the CPU_UTIL metric. The performance characteristics fall into three general regions: *IO-intensive, Balanced,* and *CPU-intensive*. The *IO-intensive* region contains applications with high IO throughput and low CPU utilization and the normalized best MSV is below 1. The *Balanced* region contains applications with medium CPU utilization and medium IO throughput and the normalized best MSV is 1 or above. The *CPU-intensive* region contains applications with high CPU utilization and low IO throughput and the best MSV is below 1. The CPU utilization or IO throughput value at which the regions separate differs depending upon a node's hardware characteristics. For a new application, CPU_UTIL measurement indicates the region the application belongs to and based on that region the MSV can be set to the predicted value.

## 5    Conclusion

Optimizing resource allocation to improve performance in Hadoop is an important area of research. Improved Hadoop performance adds value to hundreds of Hadoop deployments in commercial as well as research organizations. In this work, we explored the performance behavior of thirteen Hadoop applications that included wide range of CPU and IO characteristics. We observed that each Hadoop application has a select best MSV or MSVs for which it has the best performance. MSV is a Hadoop configuration parameter for the maximum number of map tasks that run concurrently on a tasktracker node. Additionally, there is not a single MSV that is best for all applications. Based on these findings, we developed a method to predict the best MSV. Our method uses a new Hadoop counter that measures per-map task CPU utilization. The results showed that based on the counter value the applications form three distinct regions. Each region's application has a specific range of MSV that results in its best performance. When using the region based predicted MSVs, the aggregate performance degradation is only 1 %, which is comparatively less than 7 % when using a single MSV for all applications. Furthermore, the slowdown for any application is as low as 6 % when using region based prediction compared to 20 % when using a single MSV. Thus, the low-overhead method of using metric values to predict MSV is an efficient approach for estimating the best configuration parameter value and achieving the best performance for Hadoop applications.

# References

1. Avoiding common hadoop administration issues. http://blog.cloudera.com/blog/2010/08/avoiding-common-hadoop-administration-issues
2. Hadoop poweredby. http://wiki.apache.org/hadoop/PoweredBy
3. Hadoop vaidya. http://hadoop.apache.org/docs/stable/vaidya.html
4. Ahmad, F., Lee, S., Thottethodi, M., Vijaykumar, T.N.: Puma: Purdue mapreduce benchmarks suite. Purdue University, Technical report (2012)
5. Babu, S.: Towards automatic optimization of mapreduce programs. In: Proceedings of the ACM Symposium on Cloud Computing (SOCC) (2010)
6. Chen, Y., Alspaugh, S., Katz, R.: Interactive analytical processing in big data systems: a cross-industry study of mapreduce workloads. In: Proceedings of VLDB (2012)
7. Herodotou, H., Babu, S.: Profiling, what-if analysis, and cost-based optimization of mapreduce programs. In: Proceedings of the 37th International Conference on Very Large Data Bases (VLDB) (2011)
8. Herodotou, H., Lim, H., Luo, G., Borisov, N., Dong, L., Cetin, F.B., Babu, S.: Starfish: a self-tuning system for big data analytics. In: Proceedings of Conference on Innovative Data Systems Research (CIDR) (2011)
9. Kambatla, K., Pathak, A., Pucha, H.: Towards optimizing hadoop provisioning in the cloud. In: Proceedings of the 2009 Conference on Hot topics in Cloud Computing. USENIX Association (2009)
10. Mishra, A.K., Hellerstein, J.L., Cirne, W., Das, C.R.: Towards characterizing cloud backend workloads: insights from google compute clusters. In: Proceedings of SIGMETRICS (2010)
11. Olston, C., Reed, B., Silberstein, A., Srivastava, U.: Automatic optimization of parallel dataflow programs. In: USENIX 2008 Annual Technical Conference on Annual Technical Conference (2008)
12. Polo, J., Carrera, D., Becerra, Y., Torres, J., Ayguade, E., Steinder, M., Whalley, I.:Performance-driven task co-scheduling for mapreduce environments. In: NetworkOperations and Management Symposium (NOMS), 2010. IEEE (2010)
13. Whaley, C., Petitet, A., Dongarra, J.J.: Automated empirical optimization of software and the atlas project. Parallel Comput. **27**, 3–35 (2000)
14. Zaharia, M., Borthakur, D., Sarma, J.S., Elmeleegy, K., Shenker, S., Stoica, I.: Delay scheduling: a simple technique for achieving locality and fairness in cluster scheduling. In: Proceedings of EuroSys (2010)

# A Study of SQL-on-Hadoop Systems

Yueguo Chen[1,2(✉)], Xiongpai Qin[1,2], Haoqiong Bian[1,2], Jun Chen[1,2],
Zhaoan Dong[1,2], Xiaoyong Du[1,2], Yanjie Gao[1,2], Dehai Liu[1,2],
Jiaheng Lu[1,2], and Huijie Zhang[1,2]

[1] Key Laboratory of Data Engineering and Knowledge Engineering,
MOE, Beijing, China
[2] School of Information, Renmin University of China, Beijing 100872, China
chenyueguo@gmail.com

**Abstract.** Hadoop is now the de facto standard for storing and processing big data, not only for unstructured data but also for some structured
data. As a result, providing SQL analysis functionality to the big data
resided in HDFS becomes more and more important. Hive is a pioneer
system that support SQL-like analysis to the data in HDFS. However, the
performance of Hive is not satisfactory for many applications. This leads
to the quick emergence of dozens of SQL-on-Hadoop systems that try to
support interactive SQL query processing to the data stored in HDFS.
This paper firstly gives a brief technical review on recent efforts of SQL-
on-Hadoop systems. Then we test and compare the performance of five
representative SQL-on-Hadoop systems, based on some queries selected
or derived from the TPC-DS benchmark. According to the results, we
show that such systems can benefit more from the applications of many
parallel query processing techniques that have been widely studied in the
traditional MPP analytical databases.

**Keywords:** Big data · SQL-on-Hadoop · Interactive query · Benchmark

## 1 Introduction

Since introduced by Google in 2004, MapReduce [12] has become a mainstream
technology for big data processing. Hadoop is an open-source implementation of
MR (MapReduce). It has been used in various data analytic scenarios such as web
data search, reporting and OLAP, machine learning, data mining, information
retrieval, and social network analysis [19,23]. Researchers from both industry
and academia have made much effort to improve the performance of the MR
computing paradigm in many aspects, such as optimization and indexing support of the storage layout, extension to streaming processing and iterative style
processing, optimization of join and deep analysis algorithms, scheduling strategies for multi-core CPU/GPU, easy-to-use interfaces and declarative languages
support, energy saving and security guarantee etc. As a result, Hadoop becomes
more and more mature.

© Springer International Publishing Switzerland 2014
J. Zhan et al. (Eds.): BPOE 2014, LNCS 8807, pp. 154–166, 2014.
DOI: 10.1007/978-3-319-13021-7_12

Hadoop is basically a batch-oriented tool for processing a large volume of un-structured data. However, as the underlying storage model is ignored by the Hadoop framework, when some structured layout is applied to the HDFS (Hadoop Distributed File System) data blocks, Hadoop can also handle structured data as well [17]. Apache Hive and its HiveQL language have become a SQL interface for Hadoop since introduced by Facebook in 2007.

Some researchers have compared Hadoop against RDBMSs [22], and they concluded that Hadoop is much inferior in terms of structured data processing. However, the situation has been changing recently. Traditional database vendors, startups, as well as some researchers are trying to transplant SQL functionalities onto the Hadoop platform, and providing interactive SQL query capability with a response time of seconds or even sub-seconds. If the goal is accomplished, Hadoop will be not only a batch-oriented tool for exploratory analysis and deep analysis, but also a tool for interactive ad-hoc SQL analysis of big data.

The paper firstly reviews various SQL-on-Hadoop systems from a technical point of view. Then we test and compare the performance of five representative SQL-on-Hadoop systems, based on some selected workloads from the TPC-DS benchmark. By comparing the results, strengths and limitations of the systems are analyzed. We try to identify some important factors and challenges in implementing a high performance SQL-on-Hadoop system, which could guide the efforts to improve current systems.

## 2   SQL-on-Hadoop Systems

### 2.1   Why Transplant SQL onto Hadoop

There are so many RDBMS systems in the market that support data analysis with SQL and provide real time response time. Why bother to transplant SQL onto Hadoop to provide the same function? The first reason is the cost to sale. Hadoop can run on large clusters of commodity hardware to support big data processing. SQL-on-Hadoop systems are more cost efficient than MPP options such as TeraData, Vertica, and Netezza, which need to run on expensive high end servers and don't scale out to thousands of nodes.

The second reason is the I/O bottlenecks. When the volume of data is really big, only some portion of data can be loaded into main memory, the remaining data has to be stored on disks. Spreading I/Os to a large cluster is one of merits of the MapReduce framework, which also justifies SQL-on-Hadoop systems.

The third reason is that beyond the SQL query functionality, we also need more complex analytics. SQL-on-Hadoop systems not only provide SQL query capability, but also provide machine learning and data mining functionalities, which are directly executed on the data, just like what has been done in BDAS (Berkeley Data Analytics Stack) [16]. Although RDBMSs also provide some form of in-database analytics, Hadoop-based systems however, can offer more functions, such as graph data analysis. Hadoop systems can be not only a complementary tool to RDBMSs, in some cases they can replace RDBMS systems.

The fourth reason is that, people are getting more and more interested in analysis of multi-structured data together in one place for insightful information. Hadoop has been the standard tool for unstructured data processing. If structured data processing techniques are implanted onto Hadoop, all data could be in one place. There is no need to move big data around across different tools. SQL layer will empower people who are familiar with SQL and have a big volume of data to analyze.

## 2.2 An Overview of SQL-on-Hadoop Systems

Systems coming from open source communities and startups include Hive, Stinger, Impala, Hadapt, Platfora, Jethro Data, HAWQ, CitusDB, Rainstor, MapR and Apache Drill, etc. Apache Hive and its HiveQL language have become the standard SQL interface for Hadoop since introduced by Facebook in 2007. Some works [18,20] have been done on translating SQL into MR jobs with some optimizations. Stinger [8] is an initiative of HortonWorks to make Hive much faster, so that people can run ad-hoc queries on Hadoop interactively. Impala [3] uses its own processing framework to execute queries, bypassing the inefficient MR computing model. Hadapt is the commercialized version of the HadoopDB project [9], by combining PostgreSQL and Hadoop together, it tries to retain high scalability and fault tolerance of MR while leveraging high performance of RDBMS when processing both structured and un-structured data. Platfora maintains scale-out in memory aggregates layer that roll up raw data of Hadoop. It is also a fast in memory query engine. Jethro Data [5] uses indexes to avoid full scan of the entire dataset, leading to dramatic reduction in query response time. EMC Greenplum's HAWQ [11] use various techniques to improve performance of SQL query on Hadoop, including query optimization, in memory data transferring, data placement optimization etc. Citus Datas CitusDB [2] extends HDFS in the Hadoop system by running a PostgreSQL instance on each data node, which could be accessed through a wrapper. CitusDB achieve the performance boost by leveraging structured data processing capability of PostgreSQL databases. Rainstor [7] provides compression techniques instead of a fully functional SQL-on-Hadoop system. Compression can reduce the data space used by 50X, which leads to a rapid response time. Apache Drill [4] has been established as an Apache Incubator Project, and MapR is the most involved startup in the development of Drill. Columnar storage and optimized query execution engine help to improve their query performance.

Systems from traditional database vendors include Microsoft PolyBase, TeraData SQL-H, Oracles SQL Connector for Hadoop. PolyBase [13] uses a cost based optimizer to decide whether offloading some data processing tasks onto Hadoop to achieve higher performance. TeraData SQL-H [10] and Oracles SQL Connector for Hadoop [1] enable users to run standard SQL queries on the data stored within Hadoop through the RDBMS, without moving the data into RDBMS. Systems from academia include Spark/Shark [24] and Hadoop++/ HAIL [14]. Shark [24] is a large-scale data warehouse system built on top of Spark. By using in memory data processing it achieves higher performance

than Hive. Hadoop++ and HAIL [14] improve Hadoop performance by optimizing Hadoop query plan, creating index, and co-locating data that will join
together later during data loading.

### 2.3    A Closer Look at Our Benchmarked Systems

We choose five representative systems of above for benchmarking study.

**Hive.** Apache Hive is a data warehouse software that facilitates querying and
managing big datasets in Hadoop. Hive applies structure to Hadoop data and
enables querying the data using a SQL-like language named HiveQL. Custom
mappers and reducers could be plugged in HiveQL to express complex data
processing logic. Hive supports various file formats such as char delimited text,
Sequence Files, RCFile (Row Columnar) [17], ORC (Optimized Row Columnar),
and custom SerDe (Serialization and De-Serialization). Some query optimization
strategies are applied in Hive. For example, Hive can select from Shuffle join, Map
Join, Sort Merge Join according to the data characteristics. The performance of
Hive is limited by the fact that HiveQL is translated into MR jobs to be executed
on Hadoop cluster. Some operations such as join are translated into multiple
stages of MR tasks that are executed round by round. Each task reads inputs
from disk and writes intermediate outputs back to the disk. In our benchmark,
Hive is used as the baseline to see the performance boost by the other systems.

**Stinger.** HortonWorks implements Stinger in a few steps. Firstly, it tries to
make Hive as a more suitable tool for people to perform decision support queries
by adding new features to the language and making the Hive system more like the
standard SQL model. Secondly, a cost-based query optimizer is investigated for
better query plans, and a vectorized query execution engine is applied. Thirdly,
a new columnar file format is designed for higher performance of analytic tasks.
Finally, a new runtime framework, named Tez, is introduced to reduce the Hive's
latency and throughput constraints as much as possible.

**Cloudera Impala.** Cloudera believes that Hadoop is the core of future generation of data warehouse. It involves in the competition of SQL-on-Hadoop
battle with its open sourced system Impala [3]. Impala uses its own processing
framework to execute queries, bypassing the inefficient MR computing model. It
disperses query plans instead of fitting them into a pipeline of map and reduce
jobs, thus enables parallelizing multiple stages of a query to avoid the overhead
of sort and shuffle if these operations are unnecessary. Moreover, (1) Impala
does not materialize intermediate results to disks, similar to MPP databases,
it uses in memory data transfers. (2) It avoids MR startup time by running as
a service. (3) The execution engine tries to take full advantage of the modern
hardware. Its uses the latest set of SSE (SSE4.2) instructions which can offer
tremendous speedups in some cases. (4) Impala uses LLVM (Low Level Virtual

Machine) to generate assembly code for the running queries. (5) It is aware of the disk location of blocks and is able to schedule the order to process blocks to keep all disks busy. (6) Impala is designed with performance as the top concern. Various optimization techniques are used when possible, including tight inner loops, in-lined function calls, minimal branching, better use of cache, and minimal memory usage. (7) Impala supports new columnar storages of Parquet for higher performance of query intensive workloads. According to Cloudera's benchmarking results, for purely I/O bound queries, they typically see performance gains in the range of 3-4X. For queries that require multiple MR phases or reduce-side joins in Hive, they see a higher speedup than simple single-table aggregation queries. For queries with at least one join, they have seen performance gains of 7-45X. If the data accessed by the query is resident in the cache, the speedup can be as more as 20X-90X over Hive even for simple aggregation queries [3].

**Spark and Shark.** Shark [24] is a large-scale data warehouse system built on top of Spark [25], designed to be compatible with Apache Hive. Spark provides the fine granular lineage based fault tolerance required by Shark. Shark supports Hive's query language, meta store, serialization formats, and user-defined functions. Shark can answer HiveQL queries much faster than Hive without modification to the existing data or queries. It leverages several optimization techniques, including in memory column-oriented storage layout, dynamic mid query re-planning of execution plan. These techniques allow Shark to run SQL queries up to 100X faster than Apache Hive, matching the speedups which are reported for MPP analytic databases over MR. Spark can be treated as a replacement of MR, and Shark can be treated as a replacement of Hive. The success of the Spark and Shark projects shows that by leveraging in memory data processing techniques and using careful data layouts, the Hadoop framework can achieve a fast response time to support interactive analysis.

**Presto.** Presto [6] is an interactive distributed SQL query engine that runs fast on a Hadoop Cluster. It is developed by Facebook for data analysis on petabyte-sized data warehouses. Presto is optimized for ad-hoc analysis at interactive speed by avoiding the MR. It employs a custom query and execution engine with operators designed to support SQL semantics. It uses its own query processing model. The client sends SQL to the Presto coordinator. The coordinator parses, analyzes, and plans the query execution. The scheduler wires together the execution pipeline, assigns work to nodes closest to the data, and monitors the progress. The client pulls data from output stage, which in turn pulls data from underlying stages. Presto compiles parts of the query on the fly and does all of its processing in memory. The pipelined execution model runs multiple stages at once, and streams data from one stage to the next as it becomes available. Since all processing is in memory and pipelined across the network between stages, the associated latency overhead of unnecessary I/Os is avoided. This significantly reduces the latency for many types of queries. Presto also dynamically compiles certain portions of the query plan down to byte code which lets the JVM optimize and generate native machine code. Presto can do many of the tasks that standard ANSI SQL engines can, including complex queries, aggregations, joins, left/right outer joins, sub-queries, window functions, and most of the common aggregate and scalar functions, including approximate distinct counts and approximate percentiles. The first version still lacks the ability to write results back to tables and cannot create table joins beyond a certain size. Presto supports various file formats such as Text, Sequence File, RCFile, and ORC. It scales up to a cluster of 1,000 nodes. It is 10X better than Hive/MR in terms of CPU efficiency and latency for most queries according to Facebook.

## 3  Experimental Evaluation

We select five representative SQL-on-Hadoop systems for benchmarking: Hive, Stinger, Impala, Presto and Shark. We test their performance on some selected or modified queries from the TPC-DS benchmark [21]. This section reports the results of our benchmarking tests, as well as some analysis and comparison of the performance of these systems.

### 3.1  Hardware and Software Configuration

Our experiments run on a cloud comprised of 50 physical nodes (as a part of Renda Xing Cloud[1]). Each node has a memory of 48 GB, 2 × 6 cores Intel Xeon E5645 CPU, and a disk storage of 6 TB configured with RAID 5. By using OpenStack, we are able to generate clusters of 25, 50 and 100 nodes respectively. The memory size of each virtual nodes ranges from 10 GB, 20 GB to 40 GB. A Gigabit ethernet is deployed in the clusters of our experiments.

The versions of the tested systems are listed in Table 1. The default parameters are typically applied to each benchmarked system, with some of the important parameters manually optimized for better performance.

### 3.2  Workloads

Considering that most queries of the TPC-DS benchmark are complex SQL analysis queries designed for data warehousing applications, they are not practical for many existing systems due to the computational complexity. For example, in some other studies of big data analysis systems [15,22], only simple SQL queries are executed serially to test the systems. In our study, to benchmark the systems for workloads of different complexity, we modified some queries from TPC-DS, and derive 11 queries (which are also executed serially) used in our test. Among them, some are simple single-table queries, some are the original and complex SQL queries of the TPC-DS benchmark. The applied queries include reports, ad-hoc queries, star-join queries, and complex SQL analytic queries as

---

[1] http://deke.ruc.edu.cn/yunyuyue.php

**Table 1.** Versions of tested systems

| System | Version |
|---|---|
| Apache Hive | 0.10 |
| Hortonworks Stinger | Hive 0.12 |
| Berkeley Shark | 0.7.0 |
| Cloudera Impala | 1.0.1 |
| Facebook Presto | 0.54 |

well. Here, we show two examples of them: q5Ao and q6Cgo. When naming the queries in our benchmark, 'A' indicates a single table query, 'B' indicates a join of two tables, 'C' involves 3 tables or more, 'o' indicates an 'order by' operator, and 'g' means a 'group by' operator.

```
q5Ao: select ss_store_sk as store_sk, ss_sold_date_sk as date_sk
      ss_ext_sales_price as sales_price, ss_net_profit as profit
from store_sales
where ss_ext_sales_price>20
order by  profit
limit 100;


q6Cgo: select  a.ca_state state, count(*) cnt
      from customer_address   a
      join  customer c on(a.customer_address.ca_address_sk
        = c.c_current_addr_sk)
      join  store_sales s  on(c.c_customer_sk = s.ss_customer_sk)
      join date_dim d  on(s.ss_sold_date_sk = d.d_date_sk)
      join item i  on(s.ss_item_sk = i.i_item_sk)
group by a.ca_state
having count(*) >= 10
order by cnt
limit 100;
```

We generate two datasets from TPC-DS for benchmarking the systems. One has a size of 1 TB (having a scale factor of 1,000), and the other is in 3 TB (having a scale factor of 3,000). We refer to the other benchmarks [15, 22] when choosing the dataset sizes of benchmark. The data model of TPC-DS benchmark follows a snow flake schema, with tables *store_sales* and *store_returns* as two fact tables.

### 3.3   Results

We conduct 4 groups of tests, with each having a different setting in terms of the number of virtual nodes in the cluster and the sizes of dataset. The details are listed in Table 2. In the followings, we report and compare the results from the query, system, and workload's point of views respectively[2].

---

[2] Due to the page limit, more details about the experimental settings, results, and result analysis are available at http://deke.ruc.edu.cn/sqlonhadoop.

**Table 2.** Experimental settings for 4 test groups

| Test group | No. of nodes | Data size | Results | Relative workloads per node |
|------------|--------------|-----------|---------|------------------------------|
| 1 | 25 | 1 TB | Fig. 1 | Heavy (40 GB/node) |
| 2 | 50 | 1 TB | Fig. 2 | Normal (20 GB/node) |
| 3 | 100 | 1 TB | Fig. 3 | Light (10 GB/node) |
| 4 | 100 | 3 TB | Fig. 4 | Heavy (30 GB/node) |

**Queries.** Among those 11 queries used in our benchmark, there are 6 simple queries (query 1 to query 6) that are conducted over a large fact table. Three simple join queries (query 7 to query 9) include a join operator between a dimension table and a fact table. Query 10 is a complex join query (includes both a star join and a chain join) over 5 tables. Query 11 is a star join query over multiple tables. According to the results reported in the figures, we can find that, simple queries (without join) basically perform better than those complex queries with join operations. Simple join queries basically perform better than those complex queries (query 10 and query 11). However, there are two exceptions (query 2 and query 7) which both have an 'order by' operator. A further study over these two queries show that the intermediate results (before the 'order by' operation) are very huge, which incur large cost for the 'order by' operation. Query 3 is modified from query 2 by simply applying an adjustment of the filtering condition to cause its intermediate results much less than those of query 2. As a result, we can see that query 3 performs much faster than the query 2. We should also note that queries q9* perform slightly slower than queries q5* because there are aggregation operators in queries q9*.

**Systems.** When comparing the performance of systems in different test groups, we find that impala performs much better than the others in most test cases. On the other hand, Hive often performs the worst among all the systems. This is reasonable because the other systems treat Hive as a baseline, and they try to improve their performance over Hive. Comparatively, the performance of Stinger and Presto is similar in many test cases. For Shark, we find that it performs well for light workloads, especially for simple queries over single table. However, when the workloads are heavy or when the queries are complex, Shark often fails to evaluate the queries due to many reasons such as out of memory. Note that a blank cell in the results of Figs. 1, 2, 3 and 4 represents that a system fails to execute a corresponding query.

**Workloads.** Queries are executed serially for all tested systems. We adjust the workloads of each node by varying the number of nodes and the size of dataset. For Figs. 1, 2, and 3, we actually reduce the workloads of each node by increasing the number of nodes from 25 to 100. By comparing the results of the

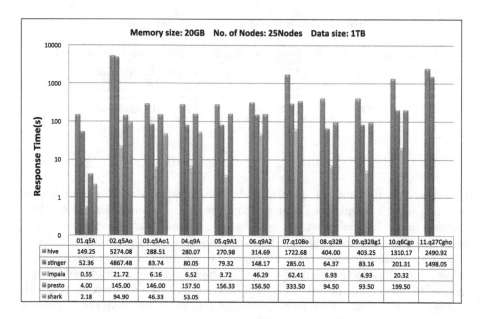

**Fig. 1.** Results on a cluster of 25 nodes for 1 TB data

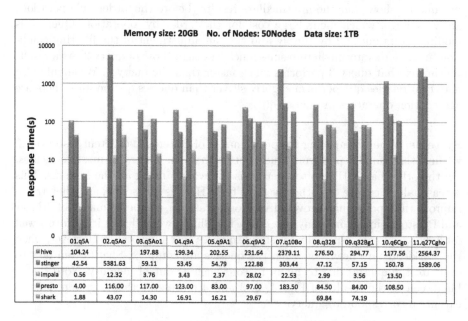

**Fig. 2.** Results on a cluster of 50 nodes for 1 TB data

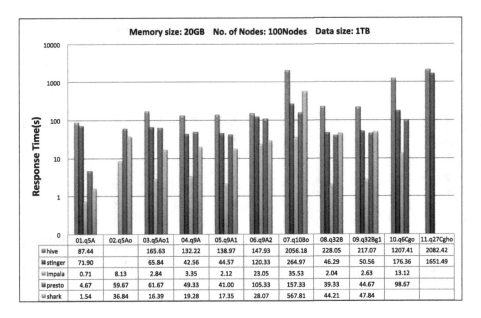

| | 01.q5A | 02.q5Ao | 03.q5Ao1 | 04.q9A | 05.q9A1 | 06.q9A2 | 07.q10Bo | 08.q32B | 09.q32Bg1 | 10.q6Cgo | 11.q27Cgho |
|---|---|---|---|---|---|---|---|---|---|---|---|
| hive | 87.44 | | 165.63 | 132.22 | 138.97 | 147.93 | 2056.18 | 228.05 | 217.07 | 1207.41 | 2082.42 |
| stinger | 71.90 | | 65.84 | 42.56 | 44.57 | 120.33 | 264.97 | 46.29 | 50.56 | 176.36 | 1651.49 |
| impala | 0.71 | 8.13 | 2.84 | 3.35 | 2.12 | 23.05 | 35.53 | 2.04 | 2.63 | 13.12 | |
| presto | 4.67 | 59.67 | 61.67 | 49.33 | 41.00 | 105.33 | 157.33 | 39.33 | 44.67 | 98.67 | |
| shark | 1.54 | 36.84 | 16.39 | 19.28 | 17.35 | 28.07 | 567.81 | 44.21 | 47.84 | | |

**Fig. 3.** Results on a cluster of 100 nodes for 1 TB data

same system in Figs. 1, 2 and 3, we can find that the enlargement of the cluster size does help to improve the performance of query evaluation. This is especially obvious for the Shark, where in Fig. 1 (with 25 nodes), only 4 queries can be successfully conducted. The number increases to 9 in Fig. 3. By comparing the corresponding numbers of the same query for the same system, we find that the speedup of query processing can hardly keep up with the rate of the cluster size enlargement. This is also reasonable because the communication cost will be increased when enlarging the number of nodes in a cluster. By keeping the number of node unchanged, and increasing the size of data (from Fig. 3 to Fig. 4), we actually increase the workloads of each node. As a result, the performance of each system drops accordingly. When the workload of each node is heavy enough (in Fig. 4), shark fails to evaluate many queries. In the meanwhile, some systems fails to evaluate query 2 and query 11.

## 3.4    Analysis

By analyzing the results of our benchmarking and referring to the system implementation, we have the following observations:

- Columnar storage is important for performance improvement, especially when the table has many columns and the query only need to access a small part of them. This is verified by Stinger (with ORCFile) over Hive, and Impala (with Parquet format and the Textfile format, whose results are not shown out).
- By discarding the MR model, the performance benefits from saving the cost of persisting the intermediate results of query processing. Impala, Shark and

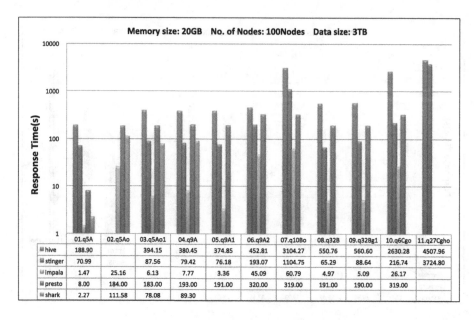

**Fig. 4.** Results on a cluster of 100 nodes for 3 TB data

Presto perform better than Hive and Stinger. They all benefit from the discarding of MR. However, the superiority decreases when the queries are complex, where the other cost will be significant.

- Techniques (e.g. distributed join processing and optimization) from MPP databases do help a lot. This is verified by the Impala system, which performs much better for join queries over two or more tables.
- Performance benefits more from the usage of large memory. Shark and Impala perform much better for small datasets. However, when the workloads increase, the available memory for each node may not be enough for implementing some join operations. This leads to many problems for memory-reliant systems such as Shark.
- Data skewness significantly affects the query performance. Systems such as Hive, Stinger and Shark are sensitive to the data skewness. Query 2 and query 7 shows that when the 'order by' operator falls in one reduce node, it will be a bottleneck of such systems.

## 4   Conclusion

In this paper, we briefly review the recent efforts of SQL-on-Hadoop systems. We test five representative systems using part of the TPC-DS benchmark. The results shows that these systems are still not efficient enough (for the interactive query processing purpose) for complex analytic queries. Even though, we find that the existing SQL-on-Hadoop systems have benefited a lot from the application of many state-of-the-art parallel query processing techniques (such as columnar storage, MPP architecture, join optimization) that have been extensively studied for many years in database community. It is expected that with more advanced parallel database techniques applied, the performance of SQL-on-Hadoop systems can be further improved. The merit of providing high performance SQL analysis functionality to the data stored in HDFS will be very attractive to many companies surfing the big data wave.

**Acknowledgements.** This work is partially supported by National 863 High-tech Program (Grant No. 2012AA011001), the National Science Foundation of China under grant No. 61472426 and No. 61170013, the Fundamental Research Funds for the Central Universities, the Research Funds of Renmin University of China No. 14XNLQ06, the Chinese National "111" Project "Attracting International Talents in Data Engineering and Knowledge Engineering Research", and the Scientific Research Foundation for the Returned Overseas Chinese Scholars, State Education Ministry.

# References

1. http://docs.oracle.com/cd/E37231_01/doc.20/e36961/sqlch.htm (2013)
2. Citusdata (2013). http://citusdata.com/docs/SQL-on-Hadoop
3. Cloudera impala (2013). http://blog.cloudera.com/blog/2012/10/cloudera-impala-real-time-queries-in-apache-hadoop-for-real/
4. Drill proposal (2013). http://wiki.apache.org/incubator/DrillProposal/
5. Jethro data (2013). http://jethrodata.com/product/
6. Presto (2013). http://prestodb.io
7. Rainstor (2013). http://rainstor.com/products/rainstor-database/
8. Stinger (2013). http://hortonworks.com/stinger/
9. Abouzeid, A., Bajda-Pawlikowski, K., Abadi, D.J., Rasin, A., Silberschatz, A.: Hadoopdb: an architectural hybrid of mapreduce and dbms technologies for analytical workloads. PVLDB **2**(1), 922–933 (2009)
10. Argyros, T.: The enterprise approach to interactive sql on hadoop data: teradata sql-h (2013). http://www.asterdata.com/blog/2013/04/the-enterprise-approach-to-interactive-SQL-on-Hadoop-data-teradata-sql-h/
11. Chang, L., Wang, Z., Ma, T., Jian, L., Ma, L., Goldshuv, A., Lonergan, L., Cohen, J., Welton, C., Sherry, G., Bhandarkar, M.: Hawq: a massively parallel processing sql engine in hadoop. In: SIGMOD Conference, pp. 1223–1234 (2014)
12. Dean, J., Ghemawat, S.: Mapreduce: simplified data processing on large clusters. In: OSDI, pp. 137–150 (2004)
13. DeWitt, D.J., Halverson, A., Nehme, R.V., Shankar, S., Aguilar-Saborit, J., Avanes, A., Flasza, M., Gramling, J.: Split query processing in polybase. In: SIGMOD Conference, pp. 1255–1266 (2013)
14. Dittrich, J., Quiané-Ruiz, J.-A., Jindal, A., Kargin, Y., Setty, V., Schad, J.: Hadoop++: making a yellow elephant run like a cheetah (without it even noticing). PVLDB **3**(1), 518–529 (2010)
15. Floratou, A., Teletia, N., DeWitt, D.J., Patel, J.M., Zhang, D.: Can the elephants handle the nosql onslaught? PVLDB **5**(12), 1712–1723 (2012)
16. Franklin, M.J.: Making sense of big data with the berkeley data analytics stack. In: SSDBM, p. 1 (2013)

17. He, Y., Lee, R., Huai, Y., Shao, Z., Jain, N., Zhang, X., Xu, Z.: Rcfile: a fast and space-efficient data placement structure in mapreduce-based warehouse systems. In: ICDE, pp. 1199–1208 (2011)
18. Iu, M.-Y., Zwaenepoel, W.: Hadooptosql: a mapreduce query optimizer. In: EuroSys, pp. 251–264 (2010)
19. Lee, K.-H., Lee, Y.-J., Choi, H., Chung, Y.D., Moon, B.: Parallel data processing with mapreduce: a survey. SIGMOD Rec. **40**(4), 11–20 (2011)
20. Lee, R., Luo, T., Huai, Y., Wang, F., He, Y., Zhang, X.:. Ysmart: yet another sql-to-mapreduce translator. In: ICDCS, pp. 25–36 (2011)
21. Nambiar, R.O., Poess, M.: The making of tpc-ds. In: VLDB, pp. 1049–1058 (2006)
22. Pavlo, A., Paulson, E., Rasin, A., Abadi, D.J., DeWitt, D.J., Madden, S., Stonebraker, M.: A comparison of approaches to large-scale data analysis. In: SIGMOD Conference, pp. 165–178 (2009)
23. Sakr, S., Liu, A., Fayoumi, A.G.: The family of mapreduce and large-scale data processing systems. ACM Comput. Surv. **46**(1), 11 (2013)
24. Xin, R.S., Rosen, J., Zaharia, M., Franklin, M.J., Shenker, S., Stoica, I.: Shark: Sql and rich analytics at scale. In: SIGMOD Conference, pp. 13–24 (2013)
25. Zaharia, M., Chowdhury, M., Das, T., Dave, A., Ma, J., McCauly, M., Franklin, M.J., Shenker, S., Stoica, I.: Resilient distributed datasets: a fault-tolerant abstraction for in-memory cluster computing. In: NSDI, pp. 15–28 (2012)

# Predoop: Preempting Reduce Task for Job Execution Accelerations

Yi Liang[1(✉)], Yufeng Wang[1], Minglu Fan[1], Chen Zhang[1],
and Yuqing Zhu[2]

[1] Department of Computer Science, Beijing University of Technology,
Beijing, China
{yliang,yufengwang,fanminglu}@bjut.edu.cn
[2] State Key Laboratory of Computer Architecture, Institute of Computing
Technology, Chinese Academy Sciences, Beijing, China
Zhuyuqing@ict.ac.cn

**Abstract.** Map/Reduce is a popular parallel processing framework for data intensive computing. For overlapping the Map task's execution phase and the Reduce task's intermediate data fetching and merging phase, existing Map/Reduce schedulers always pre-launch the Reduce task at the specific threshold where its map tasks have been launched, and this pattern incurs the occupation of the consuming resources of the reduce task during its idle time on waiting for fetching the intermediate data from map tasks. To address this issue, we propose an extension version of Hadoop map/reduce framework, called Predoop, in this paper. The basic idea of Predoop is to preempt the reduce task during its idle time and allocate the released resource to the map tasks on schedule. To achieve this goal, first, we introduce the preemptive mechanism for reduce tasks and map tasks respectively to enable Map/Reduce tasks to be preempted or resumed with correct status; second, we adopt the preempting-resuming model for the reduce task with the consideration of the progress of Reduce task data fetching & merging and the Map task execution so as to determine the timing of Reduce task preemption and resuming; third, we introduce the preemption-aware task scheduling strategy to allocate the released resources to the on-schedule Map tasks with the consideration of data locality. Experimental result demonstrates that Predoop outperforms Hadoop on various workload and the average job turnaround time can be reduced by maximum of 66.57 %.

**Keywords:** Map/Reduce · Intermediate data dependency · Preemption · Task scheduling

## 1 Introduction

Map/Reduce is a new parallel processing framework for programming the commodity computer clusters to perform the large-scale data processing [1]. The scheduling granularity for each scheduled job is on the task level in the Map/Reduce framework [1]. Once a map or a reduce task is launched, it will stay in active and occupy its allocated computing resource, such as the memory space or the CPU slots. In general, each map task always processes one data block of total job input data sets and each

© Springer International Publishing Switzerland 2014
J. Zhan et al. (Eds.): BPOE 2014, LNCS 8807, pp. 167–180, 2014.
DOI: 10.1007/978-3-319-13021-7_13

Reduce task would process the output data from all of map tasks corresponding to its processing partition. We call this data dependency between the Map task and the Reduce task as intermediate data dependency.

Due to the intermediate data dependency between map tasks and reduce tasks, existing map/reduce schedulers always pre-launch the Reduce task at the specific threshold (often 10 %) where its map tasks have been launched. Under the ideal situation, this setting can overlap the map task's execution phase and the reduce task's intermediate data fetching and merging phase [2]. However, due to the asymmetry in the completion time of map tasks of a job (because of the difference in the launching time or the uneven data processing progress), the Reduce task often stays in idle and consumes little of its occupied resources at the stage where part of its dependent map tasks have been data-fetched while others still on execution. The reduce task's idle time contributes much to the inefficient resource utilization, and hence, pulls down the efficiency of map/reduce job execution. Our examination shows that, running 20 WordCount map/reduce jobs on a 12-node cluster, the idle time of a reduce task can be, by average, 44.5 % to its total execution time and 23.3 % to its job's total execution time. The situation goes worse when the resource competition becomes more intensive (that is, more jobs in the workload) [3].

To address this performance issue, we present an extension version of Hadoop map/reduce framework, called Predoop [4]. The motivation of Predoop is to preempt the idle reduce tasks to mitigate the idle time, and allocate the resources to map tasks on schedule to accelerate the job execution. To achieve this goal, Predoop introduces a preemption model for reduce tasks to determine the time point of suspending or resuming the reduce task. Based on the preemption model, Predoop adopts a pre-emption-aware task scheduling strategy to guarantee that the on-schedule map tasks are allocated with those released resources. Further, Predoop integrates the enabling pre-emptive mechanisms for reduce tasks and map tasks to make sure that the preemption model and task scheduling are practical. The main contributions of Predoop are as follows:

(1) The preempting-resuming model for the reduce task. The definition of the time point to preempt/resume reduce task is the most fundamental factor in the reduce task preemption solution. We introduce two quantitatively estimation models—preempting model and resuming model, to determine the reduce task preempting and resuming occasion. To improve the preemption efficiency, the preempting model is designed based on the ratio of the progress of the reduce task's data fetching & merging to the map task's execution; the resuming model is design based on the ratio of the number of completing map tasks after preemption to total number of map tasks.

(2) Preemption-aware task scheduling for the reduce task preemption. Based on the preempting-resuming model, we adopt the preemption-aware task scheduling to schedule the preempted resources in high priority. We design a new scheduling strategy for preempted resources, which allocates these resources to map tasks and avoids the fragmentized execution of the map task due to its consuming resource reclaimed by the resuming reduce tasks frequently.

(3) Preemptive mechanisms for map tasks and reduce tasks. By recording the boundary of <key, value> pair processed by a map task, and the boundary of map task that has been intermediate data-fetched, the preemptive task mechanisms can assure the map and reduce task be resumed with the correct status and not losing previous works.

(4) We have conducted the performance evaluation for Predoop with two famous benchmark suites: SWIM and BigDataBench [5, 6]. The experimental results demonstrate that Predoop outperforms the native Hadoop on both the synthetic workloads (SWIM) and real world workloads (BigDataBench). It can reduce the average turnaround time of map/reduce jobs by up to 66.57 %.

The following sections are organized as follows: Sect. 2 describes the preempting-resuming model of reduce tasks; Sect. 3 present the preemption-aware task scheduling in Predoop; Sect. 4 introduces the preemptive mechanisms for the reduce task and map task respectively; Sect. 5 analyzes the experimental results; Sects. 6 and 7 present the related work, the conclusion and the future work of this paper respectively.

## 2 Preempting-Resuming Model of Reduce Tasks in Predoop

As described before, the main idea of Predoop is to preempt a reduce task during its idle time in the fetching and merging phase and allocate its released computing resources to some map tasks to be scheduled. Features of the preempting-resuming model of reduce tasks are to decide the time point to perform the preempting operation on a reduce task, and the time point to resume it and reallocate the computing resource it occupied before.

### 2.1 Preempting Model of Reduce Task

On designing the preempting model, we take two factors into consideration. One is the start point of a reduce task's idle time. It is obviously the candidate time point to preempt a reduce task. The second is the length of a reduce task's idle time. This is for that idle time of the reduce task could be too short to cover the time cost of the backfilling map task's deployment so that the benefit of utilizing the preempted resources will be overthrown. To make the most use of the limited idle time, Predoop determines the candidate time point of the reduce task preempting in an advance way by estimating the start point and the length of reduce task's idle time periodically. Once the time length is long enough, the corresponding start time will be chosen as the candidate preempting time.

For each preempting decision making, the starting point of a reduce task's idle time can be calculated out with the factor of *Remaining Fetch & Merge time* from the deciding time. The *Remaining Fetch & Merge time* can be defined as the remaining time that a reduce task needs to complete fetching and merging the intermediate data generated by the map tasks that has been completed. Due to that, in Predoop, a reduce task performs the data fetching & merging with a thread pool and fetches the intermediate data one group after another. The Remaining Fetching & Merging time is estimated as $T_{rfm}$ through the following function.

$$T_{rfm} = T_{fm} + \left\lceil \frac{N_{wait}}{k} \right\rceil \times T_{iter} \tag{1}$$

Where, $T_{fm}$ stands for the average time that a reduce task spends to fetch and merge the intermediate data from a map task; $N_{wait}$ is the number of map tasks that have been completed with their intermediate data not fetched; $k$ is the number of map tasks that a reduce task can start to fetch their intermediate data roughly at the same time and run in a single wave; $T_{iter}$ is the average time interval between two successive data fetching waves for a reduce task. In predoop, the $k$ is initially set as 1 and the $T_{iter}$ is set as the minimal remaining execution time among all active data fetching threads. As a reduce task makes progress in its execution, we dynamically update the $T_{fm}$, $k$ and $T_{iter}$ accordingly.

The figure out the length of a reduce task's idle time, we need to estimate the end time point of this time period. In predoop, the end time of a reduce task' idle time can be decided with the factor of **Remaining execution time of map task**, which the reduce task depends on, from the decision making time point. In other word, once these map tasks finish execution, the reduce task may be reactive to fetch and merge the new-generated intermediate data. The **Remaining execution time of map task** is estimated as $T_{rm}$ through the following function.

$$T_{rm} = \frac{1 - p_{mt}}{p_{mt}} \times T_{mex} \tag{2}$$

Where, $p_{mt}$ stands for the execution progress of a map task. $P_{mt}$ can be calculated out during a map task's execution according to the proportion of data that have been processed. $T_{mex}$ is the map task's total execution time since its beginning.

The preempting model of reduce task finally decides the candidate preempting time point of a reduce task according to the following condition.

$$\frac{Min\left\{T_{rm1}, T_{rm2}, \cdots, T_{rmn}\right\} - T_{rfm}}{T_{mte}} \geq D_p \tag{3}$$

Where, the set of $T_{rmi}$ $(1 \leq i \leq n)$ stands for the remaining execution time of all map tasks that the reduce task depends on; $T_{mte}$ is the average execution time of the completed map tasks that the reduce task depends on; $D_p$ is a threshold which indicates to what extend is the reduce task's idle time long enough to perform the preempting operation.

In predoop, the periodical prediction of a reduce task's preempting time will stop when a candidate time point is generated, and restart when the reduce task resumes from a preemption. To compensate the inaccuracy in the estimation, the preempting operation on a reduce task will be carried out immediately once the reduce task shifts to the idle state ahead of the candidate preempting time point. On the other hand, if the reduce task's data fetching & merging operation on its depending map tasks, which has completed on the preempting decision making time point, does not finish on the candidate preempting time point, the preempting operation of the reduce task will be postponed until all fetching & merging operations complete.

## 2.2   Resuming Model of Reduce Task

In predoop, a preempted reduce task can only be resumed when there are some map tasks that it depends on have been completed during its preemption. The resuming model determines the resuming of a reduce task only if the following two conditions are satisfied.

*Condition 1:*

$$\frac{N_{map\_c} - N_{map\_f}}{N_{map}} \geq D_r \tag{4}$$

Where, $N_{map\_c}$ stands for the number of completed map tasks that a reduce task depends on; $N_{map\_f}$ is the number of map tasks that a reduce task depends on and have completed with the generated intermediate data not fetched; $N_{map}$ is the total number of map tasks that the reduce task depends on; $D_r$ is a threshold.

*Condition 2: All map tasks allocated with the preempted computing resource of the reduce task are not in the intermediate data partition phase.*
    In a word, Condition 1 guarantees that only when the number of its depending map tasks has accumulated to be large enough, the preempted reduce task can be resumed. Condition 1 is established to prevent the frequent preempting/resuming of reduce tasks and make the reduce task fetch the intermediate data in a bundle way. Because the partition phase is the last phase of map task execution and leads to heavy disk I/O cost, Condition 2 makes the restriction that the intermediate data partitioning operation can be performed in all-or-nothing way, so as to simplify the preemption operation of map task, and make sure that the disk I/O cost caused by the data partitioning can be returned with some progress in the map task execution.

## 3   Preemption-Aware Task Scheduling in Predoop

The most distinguished feature of task scheduling in Predoop is to allocate the resources released from the preempted reduce tasks (we call them as *preempted resource*) to the on-schedule map tasks. Similar to Hadoop, task scheduling in Predoop is triggered with the 'asking for the new task' heartbeat message sent from a computing node with the available resources information enclosed. The scheduler performs the task scheduling on the preempted resources with the following three rules:

*Rule 1: The allocation of preempted resource is prior to the regular resource.*
    Where, the regular resource refers to the available resource released by the map/ reduce task that completed or failed normally.

*Rule 2:*

$$\neg \exists t (t \in PR \wedge (Aloc(t) \cap GR \neq \varphi \vee Aloc(t) \cap PR \neq \varphi)) \tag{5}$$

Where, PR stands for the preempted reduce task set in Predoop; GR stands for the non-preempted reduce task set; the function *Aloc()* can be expressed as Aloc:T –> T, where T is the total task set in Predoop. Function *Aloc(t)* defines the task set that allocated with the computing resource that task t has released when completed or suspended.

*Rule 3:*

$$\neg \exists t \left( t \in MP \wedge t \in Aloc(pr_i) \wedge t \in Aloc(pr_j), \quad pr_i \in PR, pr_j \in PR, pr_i \neq pr_j \right) \tag{6}$$

Where, MP stands for the map task set in Predoop.

Rule 1 is established for that the use of preempted computing resource is highly time sensitive (only available during the idle time of a reduce task). Rule 2 guarantees that the preempted resource can only be allocated to the map task. This is for that idea of Predoop is to preempt the reduce task only when it is idle. However, once a reduce task is allocated with the preempted resource, it may be interrupted during its data fetching when the corresponding preempted reduce task needs to be resumed and reclaim the resource. Rule 3 prevent that the resources allocated to a map task is released from multiple reduce tasks. This is to avoid the scenario where a map task 'gathers' the resource fragment from multiple suspended reduce tasks and leads to its frequent interruption because any of those reduce tasks needs to be resumed.

Based on the three rules, Predoop queues the map/reduce job in FIFO (First In First Out) way and assigns the preempted resource to map tasks with the consideration of node-level, rack-level and offSwitch-level data locality in sequence. On the other hand, among the map tasks with node-level data locality, Predoop chooses the task, that has been preempted because their consuming resources are reclaimed by the reduce tasks, in prior. This is because that the more preempted map tasks accumulated during a job's execution, the larger amount of intermediate data its reduce tasks need to fetch later, and hence increases the risk of network burst. For the regular available resources, Predoop inherits the task scheduling strategy from Hadoop.

## 4  Preemptive Task Mechanism in Predoop

In predoop, map tasks and reduce tasks are applied with different preemptive task mechanisms on their consuming resources preempted or reallocated.

As described above, the preemption of reduce task can only occur when it finishes the intermediate data fetching and merging from part of its dependent map tasks during its shuffle time. During its data fetching & merging phase, the reduce task 'pull' the intermediate data from multiple dependent map tasks in a parallel way and store them into data segments in memory or on disk according to the data size. The data segments are then merged into the larger segment and stored into the disk. Figure 1(a) shows the

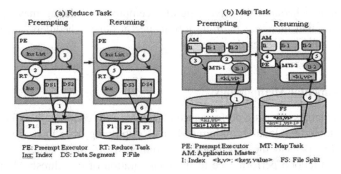

**Fig. 1.** Preemptive mechanism for reduce and map tasks

preemptive mechanism of reduce task in Predoop. In Predoop, each reducae task holds a index during it execution. The index records the dependent map tasks that have or haven't been completed yet, and the location of data segment files belonging to the reduce task. The index is updated dynamically when the reduce task is in progress. On the preemption, the reduce task first stops the updating of the index, completes the fetching and merging of data output from all completed map tasks, and flushes all data segments in memory into the disk (step 1). The reduce task then backups its index information to the Preempt Executor (step 2). Finally the Preempt Executor preempts the reduce task (actually kill the process of the reduce task) (step 3). When resuming the reduce task, the Preempt Executor first restarts the reduce task (step 4). Once restarted, the reduce task gets the index from Preempt Executor and makes clear of the map tasks to fetch the intermediate data (step 5). The reduce task then pulls the map output and merges into new data segment files (step 6).

In Predoop, the map task can only be preempted during its map phase. During the map phase, the map tasks read and process the <key, value> pair from the distributed file system HDFS in sequence. Figure 1(b) shows the preemptive mechanism of map task in Predoop. In Predoop, each map task can only be preempted at the end of each <key, value> pair and the index of the last processed <key, value> pair needs to be recorded. On preemption, the map task first finishes processing the <key, value> pair on hand. The map task then records the index of the last processed pair to its Application Master. Application Master resets the status of the map task as 'on schedule' (step 2). Finally, Application Master preempts the map task (actually kill it) via Preempt Executor resided on the same node as the map task (step 3). On resuming, the Application Master restarts the map task (step 4). The map task gets the index of last processed <key, value> pair information from Application Master (step 5). The map task then restarts the data processing from the <key, value> pair next to the last processed one.

## 5    Performance Evaluation

In this section, we present a systematic performance evaluation of Predoop. We compare the performance of Predoop to YARN (a new version of Hadoop) with the FIFO scheduler. This is for that FIFO is the fundamental of others schedulers. When porting to other scheduler, the advantage of preemptive scheduling may be amplified due to the fact that there are multiple job queues in other schedulers and concurrent preemptions can be conducted.

## 5.1 Experimental Methodology

We first conduct a systematic performance evaluation of Predoop with a diverse sets of workloads, including load-shrinking and load-amplifying workload from BigData-Bench benchmark, and the mix workload from Swim [5, 6]. Further, we study the sensitivity of Predoop performance to the configuration of two thresholds in the pre-emption model (Dp and Dr), due to the fact that various configurations result in the different occasion and frequency of reduce task preemption. Finally, we evaluate the scalability of Predoop.

Experiments are conducted in a cluster of 13 nodes. One node is dedicated as both the ResourceManager and NameNode. Each node is equipped with two Intel(R) Pentium(R) 4 cpus, 3 GB memory and one 160 GB SATA hard driver. On the YARN configuration, we configure totally 2 GB memory per node and assign 1024 MB memory for each Application Master. The HDFS block size is set as 64 MB as default.

Two benchmarks are employed. One is BigDataBench, which provides the real world ap-plication workloads with real world data sets, and we choose two single-job workloads from it: WordCount and Sort. WordCount represents the workload category that includes map/reduce job which generates small amount of intermediate data so that the reduce tasks have lighter load than map tasks (We called them load-shrinking workload). Sort represents the workload category which generates large amount of intermediate data so that reduce tasks have much heavier load than map tasks (We call them load-amplifying application). The other benchmark is SWIM. SWIM can gen-erate the synthetic workload for diverse size of Hadoop cluster according to the trace of Facebook product map/reduce platform. We add the sleep() to the map and reduce task function body so as to guarantee the execution time of map/reduce task in accordance with the heavy-tail distribution [7].

We choose Average Turnaround Time as the main evaluation metric. Average Turnaround Time is the most typical metric to reflect the efficiency of a scheduler of the system.

## 5.2 Result for the Single-Application Workloads

We first conduct the experiment on single-application workload with the input data size of 8 GB, 10 GB, 12 GB, 14 GB, 16 GB. For each job, we set the reduce task number as 8. The required memory amount of each task is set as 1024 MB as default. The thresholds $D_p$ and $D_r$ in the preemption model of reduce task are set as 20 % and 40 % respectively.

Figure 2 shows that Predoop outperform YARN on the execution of single-application workload with various data sizes. For the Sort workload, the average job turnaround time is reduced by 29.5 % on average and 49.07 % by maximum. Predoop

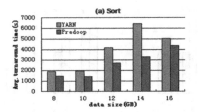 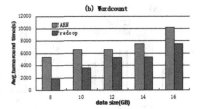

**Fig. 2.** Average turnaround time of single-application workloads

also achieves performance improvement on the load-shrinking workload, like Word-count. The average job turnaround time is reduced by 37.24 % on average and 66.57 % by maximum.

According to the statistics, due to the preemptive mechanism, Predoop minifies the reduce task's idle time by 95.86 %–99.93 % compared to YARN, and allocates the preempted resources to map tasks on schedule. This improvement contributes much to the promotion of the job turnaround time. On the other hand, the performance result shows that Predoop achieves better performance promotion on the load-shrinking workload (like Wordcount) than the load-amplifying workload. This may be for that when map tasks have heavier load and output smaller intermediate data, there will be more chances for the reduce tasks to complete one round of data fetching & merging quickly and leave more idle time to be preempted during its waiting for the next round of map task completion.

### 5.3   Results for the Mix Workloads

To evaluate the performance of Predoop in the shared map/reduce cluster, we use four mix workloads generated by SWIM. Among these mix workload, the proportion of load-amplifying job varies from 6 % to 8.7 %, which represents the typical mixture ratio in the product map/reduce clusters (like Facebook).The thresholds $D_p$ and $D_r$ in the preemption model of reduce task are also set as 20 % and 40 % respectively. To simulate the memory resource contention in the shared map/reduce cluster, we vary the memory requirement of each map and reduce task as 512 MB, 1 GB (default set in YARN), and 1.5 GB (Table 1).

Figure 3 shows that Predoop outperforms YARN on the mix workload experiment. The average job turnaround time is reduced by up to 49.85 %. According to the statistics, the drop rate of the average reduces task's idle time keeps relatively stable

**Table 1.** Characteristics of mix workloads

|                                        | Bin1  | Bin2  | Bin3  | Bin4  |
|----------------------------------------|-------|-------|-------|-------|
| Job number                             | 120   | 150   | 180   | 200   |
| Total size of map input data (GB)      | 46.66 | 64.19 | 72.32 | 82.94 |
| Total size of intermediate data (GB)   | 6     | 6.25  | 6.47  | 6.58  |
| Total size of reduce output data (GB)  | 1.36  | 1.44  | 2.26  | 7.19  |

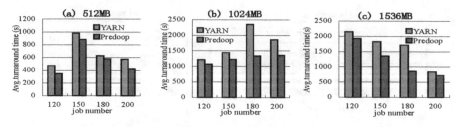

**Fig. 3.** Average turnaround time of mix workloads

under the different memory requirements (varying from 89 % to 90.7 %). However, the drop rate of the average job turnaround time goes up from 18.7 % to 25 %, with the memory requirement per task increasing (In another word, the resource contention more intensive). This is due to the fact that, with the resource contention, jobs with larger map task size may have higher risk to launch map tasks in batch. The preemptive scheduling can preempt the idle reduce tasks, contribute their occupied resource to help the on-schedule map tasks hold their required memory resource more quickly, and hence, accelerate the job completion. The statistic result shows that the performance improvement of the jobs with larger map task size contributes much to the increasing drop rate.

### 5.4    Performance Sensitivity to the Threshold Configurations

As described in Sect. 3, the preemption model of reduce tasks is designed with two threshold parameters: Dp and Dr. Performance of Predoop may be sensitive to the threshold configuration due to that these two thresholds control the occasion and frequency of reduce task preempting and resuming and may incur extra cost on the task status switch. To evaluate the performance sensitivity, we choose the mix workload

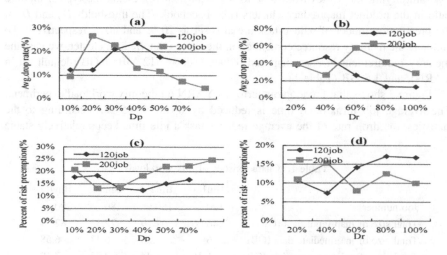

**Fig. 4.** Performance sensitivity to threshold configuration

with the job number of 120 and 200 and conduct the experiments by varying the threshold configuration. We vary Dp in preempting model as 10 %, 20 %, 30 %, 40 %, 50 %, 60 %, 70 %, and vary Dr in resuming model as 20 %, 40 %, 60 %, 80 %, 100 %.

Figure 4(a) and (b) shows the variation of average drop rate on turnaround time by Predoop. We find that for $D_p$, the best performance can be achieved when setting the threshold as 30 %–40 % for the 120-job mix workload and 20 %–30 % for the 200-job mix workload. To make it clear, we count the percent of risk preemption among all the reduce task preemption. The *risk preemption* is defined as the reduce preemption that leave the reduce task in preempting for less than 10 % of its dependent map tasks' average execution time. Figure 4(c) demonstrates that the variation of percent of risk preemption is quite in accordance with that of drop rate on trunaround time. This is due to the fact that setting the threshold too small will lead the reduce task to be preempted frequently and provide the preemption time not long enough to accommodate the efficient execution of map tasks, but only induce the extra map&reduce task start/stop cost. When setting the threshold too large, the reduce task will delay its preemption and keep it in the idle state so as to shorten the time period that map tasks consume the preempted resource.

When varying the configuration of $D_r$, we find the similar performance variation pattern as that of $D_p$ in Fig. 4(b) and (d). The cause is also similar. When the threshold is set too small, the reduce task needs to be resumed quite frequently and leaves too short preemption time. When the threshold is set too large, the reduce task will stay in the preemption even when there is enough fetching data generated, so that delay the completion of reduce tasks.

### 5.5 Scalability

We evaluate Predoop's performance scalability according to the cluster size. We generate five groups of workloads for the cluster size of 4, 6, 8, 10, 12. For each group, we generate three workloads with 120 jobs each. For each group, we calculate the average job turnaround time of the corresponding three workloads.

Figure 5 shows that with the typical synthetic workloads, Predoop outperforms YARN on diverse cluster sizes. The average turnaround time is reduced by the maximum of 46.29 %, and by the minimum of 20.98 %. According to the statistics, the average reduce task idle time is cut down by maximum of 98.37 % and by minimum of 91.28 %.

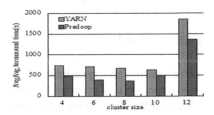

**Fig. 5.** Average turnaround time with diverse cluster size

# 6  Related Works

Many task schedulers have been proposed for map/reduce cluster over the past several years to pursue either the fairness among jobs or maximization of the job execution performance. To address the fairness issue, Hadoop, the most popular open-source implementation of Map/Reduce framework, provides three task scheduler: FIFO Scheduler, Fair Scheduler, Capacity Scheduler [3]. In [8], based on these fundamental task scheduling, the fair scheduler is improved in Hadoop by postponing the execution of head-of-queue tasks when the computing node to be allocated doesn't hold its processing data locally. Quincy introduced a min-cost flow algorithm to achieve the tradeoff between the fairness and data locality [9]. However, most fairness-centra schedulers don't adopt the preemptive mechanism and cannot prevent the long job monopolizing the system capacity or the significant resource waste.

There are several ways to maximize the job performance: (1) overlapping or subdividing some phases in a map/reduce job so as to overlap the execution of phases that utilize different resources of CPU and disk i/o. Works in [10] split the reduce task of a map/reduce job into the data copy task and the data computing task. However, it can not resolve the resource waste issue during the data copy task's idle time; (2) reducing the i/o cost by the aware of data locality or the network status. References [11, 13] present a data locality-aware and skew-aware reduce task scheduler to shorten the reduce task execution time; Maestro improves the locality of map task execution by keeping track of the data chunk and its replication location [12]. Reference [14] proposed the communication-aware placement and scheduling of map tasks and predictive load-balancing of reduce tasks so as to reduce the data i/o cost during the job execution. All the four works focus on the optimization of data i/o cost, but ignore the data dependence between map and reduce tasks; (3) remedying the outlier of map/reduce task execution particularly in the heterogonous environment. Mantri can identify the outlier in the map/reduce clusters by real-time monitoring task execution and restart the outliers on the node chosen with network awareness [15]. Though Mantri conserves some valuable work for the outlier task, the preemptive mechanism is not introduced in it. What's more, the performance optimization of the regular tasks is not Mantri's focus; (4) predicting the execution of map/reduce task and adjust the resource allocation dynamically so as to meet the SLA. ARIA conducts the job profiling and designs the map/reduce performance model to estimate the amount of resource a routinely executed job required to complete within the deadline [16]. Reference [17] adopted the preemptive mechanism for reduce task and designs the task scheduling for the fairness issue. Hence, the scheduling algorithm is totally different from that in Predoop.

# 7  Conclusion and Future Work

In this paper, we propose an extended map/reduce framework called Predoop. Predoop aims at solving the issue that the reduce task occupies the allocated resource during its idle time when waiting for the intermediate data fetching from its dependent map tasks, which lowers the job performance. Idea of Predoop is to preempt the reduce task during its idle time and allocate the released resource to the map tasks on schedule. To achieve this goal, Predoop adopts the effective preemptive mechanism for both reduce and map task, and defines the preempting-resuming model of reduce tasks with the consideration of the progress of reduce task data fetching & merging and the map task execution. Based on the preempting-resuming model, a preemptive task scheduling strategy is present to allocate the preempted resources to map tasks concerning the data locality. Experimental results demonstrate that Predoop outperforms Hadoop for the load-amplified workload, the load-shrinked workload and the mix workload. The average job turnaround time is promoted by the maximum of 66.57 %. The ongoing work includes: (1) improving the preempting-resuming model for the map/reduce job that has asymmetric processing time on multiple data elements; (2) the online adjustment of the threshold in the preemption model.

**Acknowledgements.** This work is supported by NSFC projects (Grants No. 60933003 and 61202075) and BNSF project (Grant No. 4133081).

# References

1. Chen, S., Schlosser, S.: Map-reduce meets wider varieties of applications. Technical report, IRP-TR-08-05 (2008)
2. Dean, J., Ghemawat, A.: MapReduce: simplified data processing on large clusters. In: Proceedings of USENIX Symposium on Operating Systems Design and Implementation (OSDI04), May 2004, pp. 137–150. ACM Press (2004)
3. Wang, Y.: Data dependency in map/reduce cluster. Technical report, BJUT-TR-14-01 (2014)
4. Apache Hadoop. http://hadoop.apache.org/
5. https://github.com/SWIMProjectUCB/SWIM/wiki
6. Wang, L., Zhan, J., Luo, C., Zhu, Y.: Bigdatabench: a big data benchmark suite from internet services. In: Proceedings of the 20th IEEE International Symposium on High Performance Computer Architecture (HPCA-14), pp. 21–32. ACM (2014)
7. Chen, Y., Alspaugh, S., Katz, R.: Interactive query processing in big data systems: a cross-industry study of MapReduce workloads. In: Proceedings of the 38th International Conference on Very Large Data Bases (VLDB 2012), pp. 12–23. ACM (2012)
8. Zaharia, M., Borthankur, D., Sarma, J.S.: Delay scheduling: a simple technique for achieving locality and fairness in cluster scheduling. In: Proceedings of the European Conference on Computer Systems (EuroSys'10), pp. 265–278. ACM (2010)
9. Isard, M., Prabhakaran, V., Currey, J.: Quincy: fair scheduling for distributed computing clusters. In: Proceedings of the ACM Symposium on Operating Systems Principles (SIGOPS'09), pp. 261–276. ACM Press (2009)
10. Zaharia, M., Borthakur, D., Sarma, J.S., et al.: Job scheduling for multi-user map/reduce clusters. Technical report, UCB-EECS-2009-55 (2009)
11. Hammoud, M., Rehman, M. S., Sakr, M.F.: Center-of-gravity reduce task scheduling to lower MapReduce network traffic. In: International Conference on Cloud Computing (CLOUD), pp. 49–58. IEEE (2012)
12. Ibrahim, S., Jin, H., Lu, L., et al.: Maestro: replica-aware map scheduling for MapReduce. In: International Symposium on Cluster, Cloud and Grid Computing (CCGrid), pp. 435–442. ACM/IEEE (2012)

13. Tan, J., Meng, S., Meng, X., et al.: Improving ReduceTask data locality for sequential MapReduce jobs. In: International Conference on Computer Communications (INFOCOM), pp. 1627–1635. IEEE (2013)
14. Ahmad, F., Chakradhar, S.T., Raghunathan, A., Vijaykumar, T.N.: Tarazu: optimizing MapReduce on heterogeneous clusters. In: Proceedings of the Seventeenth International Conference on Architectural Support for Programming Languages and Operating Systems (ASPLOS'12), pp. 61–74. ACM (2012)
15. Ananthanarayanan, G., Agarwal, S., Kandula, S., Greenberg, A.G., Stoica, I., Lu, Y.: Reining in the outliers in map-reduce clusters using mantri. In: Proceedings of the 9th USENIX Conference on Operating Systems Design and Implementation (OSDI'10), pp. 18–28. ACM (2010)
16. Verma, A., Cherkasova, L., Campbell, R.H.: ARIA: automatic resource inference and allocation for MapReduce environments. In: International Conference on Autonomic Computing (ICAC), pp. 235–244. ACM (2011)
17. Wang, Y., Tan, J., Yu, W.: Preemptive ReduceTask scheduling for fair and fast job completion. In: Proceedings of the 10th International Conference on Automatic Computing (ICAC-13), pp. 45–56. ACM (2013)

# Record Placement Based on Data Skew
# Using Solid State Drives

Jun Suzuki[1]([⊠]), Shivaram Venkataraman[2], Sameer Agarwal[2],
Michael Franklin[2], and Ion Stoica[2]

[1] Green Platform Research Laboratories, NEC, Kawasaki, Japan
j-suzuki@ax.jp.nec.com
[2] University of California, Berkeley, USA
{shivaram,sameerag,istoica}@eecs.berkeley.edu, franklin@cs.berkeley.edu

**Abstract.** Integrating a solid state drive (SSD) into a data store is
expected to improve its I/O performance. However, there is still a large
difference between the price of an SSD and a hard-disk drive (HDD).
One of the methods to offset the increase in cost of consisting devices
is to configure a hybrid system using both devices. In such a system,
a common method to decide the placement of data records is based on
*reference locality, i.e.,* placing the frequently accessed records in a faster
SSD. In this paper, we propose an alternative that focuses on data skew
by storing records with values that appear less often in an SSD while
those that do more in an HDD. As we will show, this enhances the
performance of fetching records using multi-dimensional indices. When
records are fetched using one of the indices targeted for optimization,
records stored in an SSD are likely be retrieved using random access,
while those stored in an HDD using sequential access. Given the method
does not rely on reference locality, its performance is stable between first
and second accesses and it provides a performance gain even when a
host memory is large enough to contain the entire working set of the
application. Our implementation and experiments show that storing just
20 % records in an SSD achieves up to 76 % of the maximum reduction
that would otherwise be obtained when all the records are stored in an
SSD.

**Keywords:** SSD · Index · Hybrid data store · Data skew

## 1  Introduction

Integrating SSDs into data stores has been a subject of much attention given
their I/O performance is higher than that of HDDs. HDDs on the other hand
have been widely used as secondary storage of data stores and their capacity has
been doubling every 18 months [6]. However, their increase in I/O bandwidth

---

J. Suzuki—Visiting scholar at University of California, Berkeley when this work was
done.

© Springer International Publishing Switzerland 2014
J. Zhan et al. (Eds.): BPOE 2014, LNCS 8807, pp. 181–193, 2014.
DOI: 10.1007/978-3-319-13021-7_14

has been slow and is currently just above 100 MB/s. SSDs on the other hand, keep doubling their bandwidth every 36 months. In addition, since they do not involve a mechanical component, their random access performance is better by more than two orders of magnitude as compared to an HDD.

Although performance of SSDs are attractive, there is still a large difference between the prices of SSDs and HDDs. In addition, their performance in sequential access is not so different as that in random access (and is even comparable when RAID is applied to HDDs). Therefore, it is reasonable to configure a hybrid system of SSDs and HDDs, and adopt a data placement method that maximizes system I/O performance.

Some of the data placement methods have been proposed for hybrid database application [1–3,9]. These methods are based on reference locality of data stored in a database. By storing frequently accessed data in an SSD, system I/O performance is improved nonlinearly to the ratio of the size of data stored in an SSD and in an HDD. Some of them also distinguish random access from sequential access, and give randomly accessed data priority to be stored in an SSD. The performance gain obtained from these methods depends on both the data access patterns and the size of the working set of application. As the hottest data is cached in the buffer pool of the host, in order to obtain a performance gain, the working set needs to contain warmly accessed data that is large enough to be spilled off from the host memory and stored in an SSD.

In this paper, we propose another data placement method among SSDs and HDDs by focusing on data skew (called *Skew-Based Data Placement, SDP,* hereafter). Data skew is known to be one of general characteristics that frequently appear in stored data [5]. SDP uses it to obtain performance enhancement nonlinear to the ratio of the size of data stored in an SSD. It provides performance enhancement from the first time when data are accessed. In addition, it provides nonlinear gain even if the host memory is large enough to contain the working set of application, and other cold data is accessed less frequently and uniformly. The application of SDP is for data that is rarely updated, – such as that for log analysis.

In SDP, the target columns for which the placement of records are optimized are given by an administrator or an overlying application. It then decides record placement so that the performance of fetching records using indices on any of those columns is enhanced. SDP configures a composite key consisting of the target columns. It then decides whether the records are to be stored in an SSD or in an HDD depending on the value of the composite key. To optimize the overall placement, data skews in all of the target columns are considered simultaneously. This process is formulated as integer linear programming (ILP) problem. With SDP, records stored in an SSD are likely to be retrieved using random access, while those stored in an HDD are likely to be fetched using sequential access. Concentrating random accesses to an SSD enables performance enhancement nonlinear to the ratio of records stored in an SSD and in an HDD.

We implemented a prototype using MySQL and evaluated it using a customer statistics table provided by an internet company, Conviva Inc. We show that by storing 20 % of records in an SSD we can achieve up to 76 % of the maximum reduction that would otherwise be obtained when all the records are stored in an SSD.

The rest of the paper is organized as follows– comparison of an SSD and an HDD is given in Sect. 2, the data skew of the table used in the prototype evaluation is discussed in Sect. 3, the proposed SDP is described in Sect. 4, the implemented prototype and its evaluation results are presented in Sect. 5, the related work is discussed in Sect. 6, and finally we conclude in Sect. 7.

## 2    Comparison of SSD and HDD

An SSD is generally composed of an SSD controller and multiple flash memory packages [4]. An SSD controller is connected to a host using some host connection interface such as SATA or PCI Express. It transforms block I/O requests from a host to read and write I/O operations to the flash memory packages. These I/O operations are parallelized among multiple packages to enhance performance of an SSD.

Table 1 compares the performance [7,8] and price [6] of a SATA SSD and an HDD. Here, we discuss read performance since that is the focus of the proposed method. Because an SSD does not contain a mechanical component and I/O requests are served in parallel among flash packages, its random access performance is better by more than two orders of magnitude than that of an HDD. On the other hand, the difference of sequential access is not as much as that of random access. When the prices of the two devices are considered, combination of RAID method and HDDs is a reasonable choice if majority of I/O traffic of application are performed in sequential access.

A data placement method of a hybrid system of SSDs and HDDs need to consider these performance characteristics. Performance gain obtained by serving random accesses using an SSD is larger than that obtained by serving sequential accesses. Therefore, to make the best of SSDs, data should be placed so that random accesses be served by SSDs while sequential accesses be done by HDDs.

**Table 1.** Comparison of SSD and HDD.

| Device | Random 4K Read IOPS | Sequential Read BW | Price |
|--------|---------------------|--------------------|-------|
| SSD | 41,000–89,000 | 500–550 MB/s | 1 $/GB |
| HDD | >91–118[a] | 125–156 MB/s | 0.05–0.07 $/GB |

[a]Calculated using read seek average time.

## 3    Skew in Data Distribution

It is widely known that in a column of a table, the numbers of entries of values appearing in that column are frequently not equal and have skew. For example, if a column is on places of residence of customers, there are many entries of big cities such as New York and Los Angeles. On the other hand, there are many other smaller cities which appear less often. Therefore, in general, there is a small number of values that often appear in a column while many others do less often.

Figure 1 shows the skew of the table we used in our prototype evaluation. It is the table on customer statistics of an Internet company, Conviva. It has 104 attributes and we investigated the skew on the values of the combination of the four columns, namely the endedFlag, customerId, country, and city. They are often used to select records in the company for data analysis. The figure shows the cumulative distribution function of the number of the occurrence of each value appearing in the table and that of its storage consumption. It shows that 90 % of values just appear in 6 % of records in the table, while 95 % of values does in 10 %. Therefore, 90 % of records are occupied by just 5 % of major values.

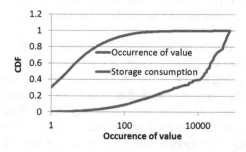

**Fig. 1.** Cumulative distribution function of the number of the occurrence of each value appearing in a table.

## 4    Optimizing Data Placement

### 4.1    Skew-Based Data Placement (SDP)

SDP uses the data skew appears in records stored in a table. Unlike conventional methods, it does not depend on the skew or locality of reference of application.

The method is intuitively illustrated using Fig. 2(a). A table which is used as an example to explain the method has two columns: customer ID and city. Although the cardinality of the customer ID is three, because of its data skew, the number of occurrence of "1" is larger than that of "2" and "3". In the same way, on the city column, the number of occurrence of "New York" is larger than that of "Berkeley", because New York is a much larger city than Berkeley.

When records, which are rows of the table, are sorted according to one column to optimize fetching records using its index, the performance of fetching records using an index on another column is not generally optimized. In a case considered here, records are fetched using either the index on the customer ID or the city. Figure 2(a) shows the order of records in the table that are sorted using a composite key of the customer ID and the city. Because the customer ID is the first column in the composite key, sorted records are clustered on it. In other words, records that share an identical value of the customer ID are continuously placed in the table. When clustered records are stored in an HDD, fetching those records are performed using sequential access. An HDD provides good performance for sequential access. On the other hand, when records are fetched using the index on the city – as the records that share an identical value of the city are separated in the table – fetching is performed using random access. When the table is stored in an HDD, fetching records that correspond to "New York" requires three seeks. Because seek time of an HDD is considerably large, the performance of fetching records using the index on the city is much worse than that of the customer ID.

SDP solves this kind of cases and enhance the performance of fetching records using indices in a multi-dimensional way. It focuses on records with less frequent values and stores them in an SSD. In the example of Fig. 2(a), when it is allowed to move up to three records from an HDD to an SSD, moving records with the customer ID of "2" and "3" reduces the number of seeks to fetch the records with "New York" from three to one. This means that by moving 33 % of records of the table, the number of seeks is reduced by 66 %. In this way, by moving records with less frequent values, SDP reduces the number of seeks made to an HDD nonlinearly to the ratio of the size of records stored in an SSD and an HDD.

Figure 2(b) schematically shows the reduction of the cardinality of the composite key of records stored in an HDD, The horizontal and vertical axises correspond to the columns consisting the composite key. Although the number of axises or columns in the composite key is two for the explanation purpose, SDP is not limited to it. Each square in Fig. 2(b) corresponds to a possible value of the composite key which are combination of values in each column. The colored squares represent that there are records with the corresponding value. By storing less frequent values of the composite key in an SSD, large number of colored

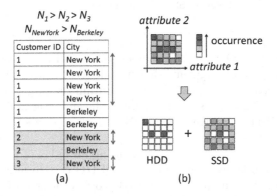

**Fig. 2.** Enhancement of fetching records using data skew.

squares moves to an SSD. As a result, an HDD stores a small number of colored values that has many entries. This results in the nonlinear reduction of the number of seeks in fetching records using indices of the columns in the composite key except for the first column.

## 4.2  Optimization Formulation

In this subsection, the optimization of data placement between an SSD and an HDD is formulated. It is supposed that data are stored in a single table, and records are retrieved using either of multiple indices. Therefore, the motivation of the data placement is to enhance the performance of fetching records using indices in multi-dimensional way.

It is also shown that the optimization of data placement is formulated as integer linear programming. I/O cost to fetch records is the optimized function under the constraint of SSD resources. Because integer linear programming is known as an NP-hard problem, we used greedy method in our prototype described in Sect. 5 to perform optimization calculation.

**Table 2.** Variables to formulate I/O cost

| Variable | Explanation |
|---|---|
| $N_i$ | Cardinality of column $c_i$ |
| $K_i$ | Cardinality of combinatorial column $(c_1, c_2, ..., c_i)$ |
| $F_i$ | The number of fragmentation of column $c_i$ in HDD |
| $B_{SSD}, B_{HDD}$ | Bandwidth of SSD or HDD |
| $S_{SSD}, S_{HDD}$ | Size of data stored in SSD or HDD |
| $T_{seek}$ | Average seek time of HDD |
| $y_i(x_1, x_2, ..., x_i)$ | Whether records with $(x_1, x_2, ..., x_i)$ are stored in SSD |
| $s(x_1, x_2, ..., x_n)$ | Total size of records with combinatorial values $(x_1, x_2, ..., x_n)$ |
| $C_{SSD}$ | Constraint of SSD consumption |

To formulate the I/O cost, the variables shown in Table 2 are introduced. $F_i$ represents the number of fragments that include records with the same value on column $c_i$ in an HDD. $i$ represents the order within the target columns that are optimized for fetching records. In the example shown in Fig. 2, if the whole table is stored in an HDD, $F_2$ is five because "New York" appears in three fragments while "Berkeley" does in two.

The average I/O cost to fetch records using an index of $c_i$ by specifying a value in the column is given by dividing total I/O cost to fetch all records by the cardinality of the column. The total I/O cost is the sum of the I/O cost to fetch records reside in both an SSD and in an HDD. We ignored the latency of I/O requests other than the seek time of an HDD, because our interest is the difference of the cost between the two devices. Then, the average I/O cost is described as

$$\frac{1}{N_i}\left(T_{seek}F_i + \frac{S_{HDD}}{B_{HDD}} + \frac{S_{SSD}}{B_{SSD}}\right). \tag{1}$$

When the size of an SSD that can be used to store records of a table is given as the constraint of the data placement optimization, the second and third terms in formula (1) are constant. Then, only the first term varies depending on the placement of records. Therefore, we formulate the I/O cost to be optimized by linearly combining the costs of each target column as

$$\sum_i R_i \frac{F_i}{N_i} \tag{2}$$

Here, $R_i$ adjusts the relative importance among the target columns, and $T_{seek}$ is included into it. We introduce two parameters, $y_i$ and $s$; $y_i(x_1, x_2, ..., x_i) \in \{0, 1\}$ denotes whether records with the combination of the values in target columns of $(x_1, x_2, ..., x_i)$ are stored in an SSD; $s(x_1, x_2, ..., x_n)$ denotes the total size of the records with the combination of the value of $(x_1, x_2, ..., x_n)$. $n$ denotes the number of columns to be optimized and $1 \le i \le n$. In SDP, records which share the same combination of the value $(x_1, x_2, ..., x_n)$ are stored either in an SSD or an HDD. Then, the next relation holds for $y_i$ and $y_{i+1}$.

$$y_{i+1}(x_1, x_2, ..., x_i, x_{i+1}) - y_i(x_1, x_2, ..., x_i) \geqq 0 \tag{3}$$

That is, for $y_i(k_1, k_2, ..., k_i)$ of the specific combination of constants of $(k_1, k_2, ..., k_i)$ to be one, $y_{i+1}(k_1, k_2, ..., k_i, x_{i+1})$ needs to be one for all possible value of $x_{i+1}$. Then, the number of fragments $F_i$ in an HDD is described as

$$F_i = K_i - \sum_{x_1}\sum_{x_2}...\sum_{x_i} y_i(x_1, x_2, ..., x_i). \tag{4}$$

The constraint of the consumption of SSD resources is described as

$$\sum_{x_1}\sum_{x_2}...\sum_{x_n} s(x_1, x_2, ..., x_n)y_n(x_1, x_2, ..., x_n) \leq C_{SSD}. \tag{5}$$

Substituting formula (4) into formula (2) gives the I/O cost that are described with variables $y_i(x_1, x_2, ..., x_i)$. Then, calculating the combination of $y_i(x_1, x_2, ..., x_i)$ which minimizes the I/O cost under the constraints of formulas (3) and (5) is an integer linear programming since $y_i(x_1, x_2, ..., x_i) \in \{0, 1\}$. An integer linear programming is known as an NP-hard problem.

## 5    Prototype Evaluation

### 5.1    Implementation

To evaluate SDP, we implemented a prototype shown in Fig. 3 using MySQL. It is consisted of two layers: data placement optimizer and MySQL. The data placement optimizer optimizes the placement of records in an original table between an SSD an HDD. An application program is supposed to create individual index on each of the target columns and use either of them to fetch records by specifying value on that column.

The data placement optimizer takes the statistics on the occurrence of values in the target columns. It then decides whether records with each combination of values are stored in an SSD or in an HDD. It uses the data placement optimization explained in Sect. 4 in the constraint of the SSD resources given by an administrative interface. It also creates an additional column, $ssd\_flag \in \{0, 1\}$, to an original table. The flag denotes whether a record containing a flag is stored in an SSD or not.

Because the calculation of the data placement optimization is NP-hard, greedy method is used. It continued choosing the next combination of values that reduced the I/O cost per SSD consumption described by formula (2) most until the total consumption of an SSD reached the given constraint.

The range partition function of MySQL 5.6 was used to divide storing records between an SSD and an HDD. Records with ssd_flag of "1" were stored in an partition made in the directory where an SSD was mounted while ones with ssd_flag of "0" were stored in the HDD directory.

In the evaluation, SSD resources consumed by the indices and the ssd_flag column was ignored because it was small compared to the original data table that had 104 columns. We used a single host with two 8-core xeon CPUs and 128-GB memory. It contained both a SATA SSD (Intel 520 series) and a SATA HDD (Seagate ST91000640NS Constellation.2) that were used for SDP evaluation.

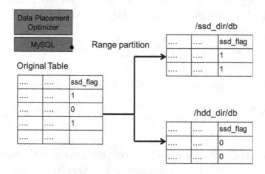

**Fig. 3.** Prototype implementation.

## 5.2  Evaluation

We performed two kinds of experiments. In the first experiment, the columns that were frequently used to select records in the data analysis in an Internet company, Conviva were selected as the target of optimization. However, because these columns were correlated, records that shared the same value in the target column were not fragmented so much and the gain of the performance by SDP was limited. Therefore, in the second experiment, different columns were selected and the size of records was halved to artificially increase the affect of seek of an HDD to the system performance. The table used in the experiment is the same as the one analyzed in Sect. 3 that had 1,062,701 records, each record consisted of 104 columns, and the size of each record was about 2 KB. For the simplisity of evaluations, columns that were not in the interests were defined as a single large blob column.

In the first experiment, data placement was optimized on the columns that were frequently used for analysis. Table 3 shows the cardinality of each of the four columns chosen to be optimized, namely endedFlag, customerId, country, and city. It also shows the cardinality of their combination.

In SDP, records in an SSD and an HDD are sorted using a composite key that is consisted of the target columns for optimization. Except for the first column in the composite key, how much the records sharing the same value in a column are fragmented depends on the cardinality of its preceding columns. Therefore, to minimize the fragmentation of records in an HDD, the order of columns in a composite key is decided in ascending order of their cardinality; in the first experiment, the composite key was (endedFlag, customerId, country, city).

**Table 3.** Cardinality of each and composite column used in the experiments.

| Key | Cardinality |
|---|---|
| endedFlag | 2 |
| customerId | 7 |
| country | 179 |
| city | 6086 |
| connType | 12 |
| isp | 130 |
| (endedFlag, customerId) | 12 |
| (endedFlag, customerId, country) | 462 |
| (endedFlag, customerId, country, city) | 10813 |
| (city, customerId, connType, isp) | 19587 |

The performance of SDP was evaluated by the average cost of fetching records selecting single value on the focused column. Figure 4 shows the evaluated performance on the select queries on city, which was the last column in the composite key. The results show that by storing 20 % of records in an SSD, the average response time of the queries was reduced by 52 % of the maximum reduction which was obtained when all records were stored in an SSD. On the other hand, the reduction of the response time of 90th percentile was 75 % of the maximum reduction.

The results of city in Fig. 4 also shows that the response time of the 90th percentile is faster than that of the average. This is caused by the data skew in the city column; a small number of major cities in the column largely increases the average response time. In addition, the reduction of the response time of the 90th percentile is larger than that of the average. We consider this is due to the difference of the ratio of the seek and read time between major and minor values. On major values, because there are many records, the ratio of read time is larger than that of minor values. Therefore the performance gain by reducing the number of seeks could be suppressed for major values.

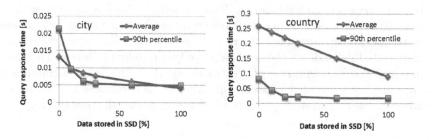

**Fig. 4.** Query response time on city and country in the first experiment.

In the first experiment, however, the nonlinear performance enhancement was not obtained in the preceding columns to the city in the composite key. Figure 4 also shows the performance of the queries on country, which was the third column in the composite key. Although the response time of the 90th percentile is slightly improved nonlinearly, the reduction of the average response time is linear. This shows that on the preceding columns including the country, the bottleneck of the performance to fetch records is the bandwidth of the sequential read of the HDD. Therefore, by storing some of records in an SSD, the performance is improved linearly and its slope is decided by the difference of the bandwidth of the SSD and the HDD.

In the second experiment, the different columns were chosen from the first experiment and the length of the records was halved to increase the affect of the seek of an HDD. The target columns for optimization were customerId, connType, isp, and city, and the composite key was configured using these columns in this order. However, also in this case, the bottleneck for all the column except the city was the read bandwidth of the HDD. Therefore, the order of the column was changed and the composite key of (city, customerId, connType, isp) was also tried. In this case, the bottleneck for all of the column was the seek time of the HDD. The cardinality of regarding columns are shown in Table 3.

Figure 5(a) shows the query response time on city when (customerId, connType, isp, city) was used as the composite key. Because the affect of the seek time is increased from one in the first experiment, by storing 20 % of records, 76 % of maximum reduction that is obtained when all records are stored in the SSD is obtained. Figure 5(b) shows the cumulative distribution function of the

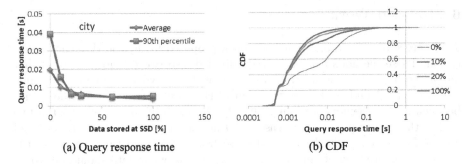

**Fig. 5.** (a) Query response time and (b) its cumulative distribution function (CDF) on city in the second experiment. Legend in CDF is the ratio of records stored in SSD.

query response time on the city. When the ratio of the records stored in an SSD is increased, the response time rapidly approached to the performance with the maximum reduction of response time.

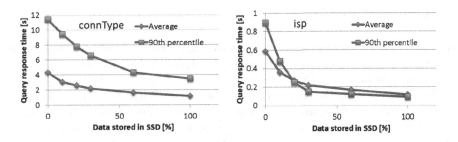

**Fig. 6.** Query response time on connType and isp in the second experiment.

Next, when (city, customerId, connType, isp) was used as the composite key, the nonlinear reduction of query response time was obtained for all of the target columns. Figure 6 shows the query response time on the connType which was third in the composite key and the isp which was the last. When 20 % of records are stored in an SSD, the average response time of the select queries is reduced by 55 % and 68 % of the maximum reduction for the connType and the isp, respectively.

These evaluation results showed that SDP provides nonlinear reduction of query response time against the ratio of the size of records stored in an SSD and an HDD. It also reduces the response time in multi-dimensional way. How large the nonlinear performance gain is depends on several factors such as the cardinality, the correlation, and the record size of the data stored in a system.

## 6    Related Work

There are several methods that have been proposed for hybrid database application. These methods can be broadly divided into two categories – ones that handle both an SSD and an HDD as different devices which configure the same storage tier [1] and others that use an SSD as an intermediate cache device between a host memory and an HDD [2,3,9]. The difference between the two is that the former provides increased I/O performance when data are accessed for the first time, while the latter does so from the second time. However, the former needs profiling of application to decide as to what data should be stored in an SSD which the latter does not.

All of these methods are based on reference locality of data stored in a database. By storing frequently accessed data in an SSD, system I/O performance is improved nonlinearly to the ratio of the size of data stored in an SSD and in an HDD. On the other hand, in this paper, we propose an alternative that focuses on data skew by storing records with values that appear less often in an SSD while those that do more in an HDD.

## 7    Conclusions

In this paper, we proposed SDP to enhance the performance of fetching records stored in a hybrid data store of SSDs and HDDs using indices on different columns. It is based on the data skew of stored data, and provides nonlinear performance gain to the ratio of records stored in SSDs and HDDs. Because SDP uses the data skew, unlike caching, it provides stable performance enhancement between first and second data accesses. It can also enhance performance even when a system memory is large enough to contain the working set of application. The evaluation of the implemented prototype using the data from the internet company showed that the performance of fetching records using different indices are simultaneously enhanced. By storing 20 % of records, up to 76 % of the maximum reduction of query response when all records are stored in an SSD is obtained. Our future work includes comparing the performance of SDP and others that are based on reference locality, and evaluation using real queries that are used in the data analysis.

## References

1. Canim, M., Mihaila, G.A., Bhattacharjee, B., Ross, K.A., Lang, C.A.: An object placement advisor for DB2 using solid state storage. In: VLDB, pp. 1318–1329 (2009)
2. Canim, M., Mihaila, G.A., Bhattacharjee, B., Ross, K.A., Lang, C.A.: SSD buffer-pool extensions for database systems. In: VLDB, pp. 1435–1446 (2010)
3. Do, J., Zhang, D., Patel, J.M., DeWitt, D.J., Naughton, J.F., Halverson, A.: Turbocharging DBMS buffer pool using SSDs. In: SIGMOD (2011)

4. Agrawal, N., Prabhakaran, V., Wobber, T., Davis, J.D., Manasse, M., Panigrahy, R.: Design tradeoffs for SSD performance. In: 2008 USENIX Annual Technical Conference (ATC'08), pp. 57–70 (2008)
5. Walton, C.B., Dale, A.G., Jenevein, R.M.: A taxonomy and performance model of data skew effects in parallel joins. In: VLDB, pp. 537–548 (1991)
6. Stoica, I.: Warehouse-Scale Computing and the BDAS Stack. http://ampcamp. berkeley.edu/amp-camp-one-berkeley-2012/
7. Intel SSD Product Comparison. http://www.intel.com/content/www/us/en/ solid-state-drives/solid-state-drives-ssd.html
8. Seagate Desktop HDD. http://www.seagate.com.edgekey.net/staticfiles/docs/pdf/ datasheet/disc/desktop-hdd-data-sheet-ds1770-1-1212us.pdf
9. Liu, X., Salem, K.: Hybrid storage management for database systems. In: VLDB, pp. 541–552 (2013)

# Efficient HTTP Based I/O on Very Large Datasets for High Performance Computing with the Libdavix Library

Adrien Devresse$^{(\boxtimes)}$ and Fabrizio Furano$^{(\boxtimes)}$

CERN, European Organization for Nuclear Research, Geneva, Switzerland
{adrien.devresse,furano}@cern.ch
http://cern.ch

**Abstract.** Remote data access for data analysis in high performance computing is commonly done with specialized data access protocols and storage systems. These protocols are highly optimized for high throughput on very large datasets, multi-streams, high availability, low latency and efficient parallel I/O. The purpose of this paper is to describe how we have adapted a generic protocol, the Hyper Text Transport Protocol (HTTP) to make it a competitive alternative for high performance I/O and data analysis applications in a global computing grid: the Worldwide LHC Computing Grid. In this work, we first analyze the design differences between the HTTP protocol and the most common high performance I/O protocols, pointing out the main performance weaknesses of HTTP. Then, we describe in detail how we solved these issues. Our solutions have been implemented in a toolkit called davix, available through several recent Linux distributions.

Finally, we describe the results of our benchmarks where we compare the performance of davix against a HPC specific protocol for a data analysis use case.

**Keywords:** High performance computing · Big data · HTTP protocol · Data access · Performance optimization · I/O

## 1 Introduction

The Hyper Text Transport Protocol (HTTP) [9] is today undoubtedly one of the most prevalent protocols on the internet.

Initially created by Tim Berners-Lee for the World Wide Web at CERN in 1990 [8], HTTP is today much more than a simple protocol dedicated to HTML web page transport. The extensions of HTTP like WebDav [21] or CalDav [15], the HTTP based protocols like UPnP or SOAP [10] and the RESTful [17] architecture for Web Services have transformed HTTP into an universal versatile application layer protocol for the internet. The recent emergence of cloud computing [13] and the popularization of big data storage [31] based on RESTful Web services have definitively proved the universalism of HTTP.

© Springer International Publishing Switzerland 2014
J. Zhan et al. (Eds.): BPOE 2014, LNCS 8807, pp. 194–205, 2014.
DOI: 10.1007/978-3-319-13021-7_15

Today, HTTP is the foundation for interactions with commercial cloud storage providers like Amazon Simple Storage Service [27] or with Open Source Cloud Storage [5] systems like OpenStack Swift [29] using REST API like S3 [26] or CDMI [36]. HTTP is fully accepted as data transfer and data manipulation protocol in NoSQL databases and distributed storage systems in the Web World. The association of REST APIs with the HTTP protocol usage offers a simple, standard, extensible, portable alternative to the legacy data access and file manipulation protocols or to the proprietary protocol of most distributed file systems.

However, the popularity of HTTP was still not penetrating the High Performance Computing world. HPC[1] data access have highly specific and strict requirements: very large data manipulation, low-latency, high-throughput, high-availability, highly parallel I/O and high-scalability. For these reasons, those use cases traditionally rely on highly specific systems and protocols adapted to such constrains. The IBM GPFS [35] protocol, the Lustre parallel distributed file system protocol, the Hadoop HDFS [42] streaming protocol, the gridFTP protocol [11] or the XRootD protocol [31] are widely used in super computing and grid computing environments.

The focus of this work is to be able to make the HPC world benefit of all the momentum coming from the HTTP Ecosystem, like the RESTful and Cloud Storage services, by creating a high performance I/O layer based on the HTTP protocol, compatible with standard services and competitive with the HPC I/O specific protocols.

To achieve this, we have created *libdavix* [1,38], an I/O layer implementation optimized for data analysis and HPC I/O in distributed environments.

## 2 Background and Related Work

### 2.1 HTTP as a Data Management and Data Transfer Protocol

The stateless nature of HTTP, associated with the atomic nature of its primitives, provides a simple, reliable and powerful consistency model in distributed environment. The success and the major diffusion of the RESTful architecture introduced by Roy Fielding [17] for the Web World illustrate perfectly this [34].

The HTTP PUT method provides an object level idempotent write operation that can be used for an atomic resource creation or a resource content update. The basic HTTP GET method gives a safe, cacheable, atomic and idempotent Object level read operation and can be used to access and retrieve safely a remote resource. These two methods, associated with the HTTP DELETE method, are enough to satisfy the four basic functions Create, Retrieve, Update, Delete (CRUD) [32] of any basic persistent storage system [6].

The properties of the HTTP protocol make it suitable for data transfer in a distributed environment and easily justify the emergence of persistency and data storage solutions using RESTful interfaces. It is the case for instance of the

---

[1] High Performance Computing.

NoSQL database couchdb [3], of the NoSQL key-store Ryak [2], of the distributed file system HDFS with httpFS [42], of the Amazon Simple Storage Service (Amazon S3) [27] or of any similar RESTful Object Storage service.

Again, the universalism of HTTP and the quality of its ecosystem associated with its simple and flexible design, makes it a first quality choice for a generic data transfer protocol today.

## 2.2 Efficient Parallel I/O Operations

Intensive data analysis applications requires high degree of I/O parallelism, robustness over large transfers and low I/O latency. A high energy physics application typically processes in parallel a very large number of events from different files located in large distributed data stores, triggering a significant number of parallel I/O operations.

For such use cases in, the grid computing models in the HEP community use a mix of I/O frameworks for HPC[2] data access, the XRootD framework [16], the GridFTP protocol [11] with the Globus toolkit [18], HDFS of Hadoop [42] and IBM GPFS [35].

All these frameworks are highly optimized for parallel access, high throughput and efficient I/O scheduling of multiple requests. The XRootD framework implements its own I/O scheduler, it supports parallel asynchronous data access on top of its own I/O multiplexing.

The GridFTPv2 protocol has separated control and data channels and supports multiple data streams from different data sources on top of TCP or UDP.

The HDFS architecture is specially designed for large file storage, high throughput, hot file replications and data streams from multiple DataNodes.

To explore the possibilities for a solution based on HTTP, we defined a set of criteria to meet.

- Efficiency for large data transport and parallel I/O.
- Compatibility with existing network infrastructure and services.
- Low I/O latency: avoid useless handshakes, useless reconnections and redirections.

The original design of HTTP did not match very well these points.

The HTTP 1.0 standard recommends the usage of one TCP connection per request to the server. This approach has been already proven inefficient due to the TCP slow start mechanism [37]. Executing a HEP[3] data analysis work-flow with a very large number of parallel small sized requests in such conditions would lead to a major performance penalty.

To mitigate the effect of this behaviour, HTTP 1.1 introduced the persistent connection support with the *KeepAlive* mechanism and the support of request pipelining over the same connection [34] (Fig. 1).

---

[2] High performance computing.
[3] High energy physics.

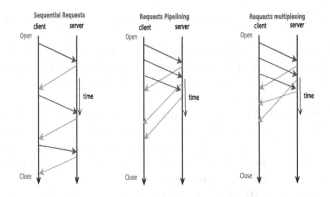

**Fig. 1.** HTTP Request pipelining and Request Multiplexing

However, HTTP pipelining suffers of several problems. Contrary to a proto-col supporting modern multiplexing, the HTTP pipelining suffers of the HOL[4] problem.

The HTTP standard specifies that the treatment of a group of pipelined HTTP requests has to be processed in order. With such a requirement, any request pipelined suffering of a delay will cause a delay for all the following requests [28]. This is a unacceptable performance penalty in case of parallel I/O requests with different sizes.

The HTTP pipelining also suffers from other problems. It suffers of side effects with the TCP's nagle algorithm [22]. It often suffers of performance degradation due to aggressive pipeline interruptions with some servers imple-mentations and due to the fact that the pipelining is not respected by most of the proxy servers. For these reasons, most of the current web browsers (Chrome, Firefox and Internet explorer) web browser disable or do not support the HTTP pipelining mechanism.

To resolve these problems inherent to HTTP 1.1, several proposal have been made:

- the **SPDY protocol** is "an application-layer protocol for transporting con-tent over the web, designed specifically for minimal latency". SPDY acts as a session layer between HTTP and TCP. It supports multiplexing, prioritization and header compression [7]. SPDY is currently the most mature implementa-tion of multi-plexing for HTTP and supported by a majority of modern web browsers.
- **HTTP over SCTP** proposes to use the SCTP multi-homing and multi-plexing features with HTTP [33]. SCTP is a acronym for Stream Control Transmission Protocol, it provides a message-based alternative to TCP.
- **WebMux and HTTP-NG** proposes the usage of the MUX protocol as session protocol to provide multi-plexing for HTTP [20,30].

---

[4] Head of Line.

None of these approaches are considered acceptable for a High performance I/O usage.

The SPDY protocol explicitly enforces the usage of SSL/TLS for protocol negotiation purpose. TLS introduces a negative performance impact for big data transfers [14] and introduces a handshake latency that can not be mandatory in High performance computing.

The HTTP over SCTP proposal implies naturally to replace the TCP protocol by SCTP. Like any other protocol aiming to replace the level 4 of the OSI model, the SCTP protocol triggers several concerns about the compatibility with the existing network architecture, about the NAT-traversals capability and about the support in old operating systems.

The MUX, renamed WebMUX protocol, defined in 1998, focuses on an object oriented approach with in mind the support for technology like RMI[5], DCOM[6] or CORBA[7] which is not our use cases. Moreover, it has currently not been implemented in any major HTTP server.

To satisfy our use cases, we adopted and implemented a different approach into *libdavix* [1,38]. We use a hybrid solution based on a dynamic connection pool with a thread-safe query dispatch system and a session recycling mechanism (See Fig. 2).

Associated with the pool, we enforce an aggressive usage of the HTTP KeepAlive feature, *libdavix* to maximize the re-utilization of the TCP connections and to minimize the effect of the TCP slow start.

This method gives several benefits compared to previously quoted solutions. First, it is fully compatibility with the standard HTTP 1.1 and with existing services and infrastructure.

Second, we supports efficient parallel request execution for repetitive I/O operations without suffering of the problems that are specific to the classical HTTP pipelining. nor necessitating a protocol modification to support multiplexing.

This dispatch approach is particularly adapted to a HPC data-analysis workflow: a repetitive concurrent access to a limited set of hosts exploits at the best the session recycling and maximizes the lifetime of the TCP connections. However, contrary to a pure multi-plexing solution that aims to the usage of one TCP connection per host, our approach uses a connection pool whose size is proportional to the level of concurrency. Consequently, an important degree of concurrency can result in a more important server load compared to a multiplexed solution like spdy due to the number of connections allocated per client.

However, this is not a big issue for our HPC use case, the support of "vectored queries" of *libdavix* explained in Sect. 2.3 reducing significantly the number of simultaneous concurrent requests.

---

[5] Remote Method Invocation.

[6] Distributed Component Object Model.

[7] Common Object Request Broker Architecture.

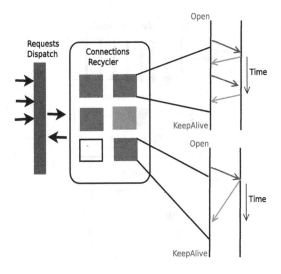

**Fig. 2.** Dynamic connection pool with a thread-safe query dispatch

## 2.3 Scalable Random-Access I/O for Partial File Access with HTTP

Very large data sets in distributed environments are generally partitioned into separated data subset objects distributed in several storage nodes.

In High Energy Physics, a data set generally contains a important number of particle events decomposed in ROOT [4] *TTrees* of events and stored in different compressed files.

This approach allows an easy data distribution and simplify data replication. It facilitates the partitioning the data in a distributed environment like in the World Wide LHC Computing Grid (WLCG) [23]. At the same time, a data analysis with this data model is I/O intensive and generates a very large number of individual data accesses operation to the storage. In order to extract a set of specific events spread in different remote data sets, a HEP Application needs to read a large number of small segments of data in different remote objects.

To reduce the number of I/O requests, and, hence, the impact of latency with such patterns, high performance computing I/O protocols implement Data Sieving algorithms, two phase I/O algorithms [39] or sliding window buffering algorithms.

To the best of our knowledge, no nowadays HTTP I/O toolkit before davix implemented similar I/O optimization strategies.

We implemented in *libdavix* a support for vectored packed operations with random position based on the multi-range feature of the HTTP 1.1 protocol (Fig. 3).

This feature allows to gather and pack a large number of fragmented random I/O requests directly in the ROOT [4] I/O framework via the TTreeCache [41] in a large vectored query. Subsequently, this query is processed by *libdavix* as

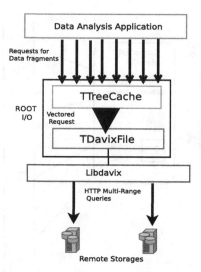

**Fig. 3.** Vectored I/O requests support in the ROOT framework associated with Lib-Davix

one atomic remote I/O query. This approach reduces drastically the number of remote network I/O operations and offers the advantage to reduce the necessity of parallel I/O operations and thus virtually eliminates the need for I/O multiplexing.

### 2.4   Multi-stream and Multiple Replicas I/O Operations

In grid computing, the unavailability of an input data required by a job is often the main cause of failure. Such situation leads to a redistribution of the data and a rescheduling of the job with a substantial performance impact on the execution time.

In an attempt to solve this issue, the XRootD protocol [16] supports a federation mechanism to offer a more resilient access to a distributed resources. XRootD data servers can be federated hierarchically into a global virtual namespace. In case of unavailability of a resources in the closest data repository, the XRootD federation mechanism will locate a second available replica of this resource and redirect the client there.

HTTP alone does not support this feature. A classical HTTP access to an unavailable resource or a offline server will result in a I/O error. To improve the resiliency of the data layer, davix implements natively a support for the Metalink [25] standard file format coupled to a fail-over and filtering mechanism for the offline replicas of a resource.

A Metalink file is a standardized XML [12] file containing several elements of meta-data information about an online resource: name, size, checksum, signature and location of the replicas of the resource. A Metalink file is a resource description and a set of ordered pointers to this resource.

Davix can use the Metalink information to apply two strategies:

- The **"fail-over" strategy** (default). In the case a resource is not available, davix try seamlessly to obtain the Metalink associated with this resource. Then *libdavix* will try to access one per one the available replicas of this resource until being able to access the requested data on one of the available replica of the resource. This approach improves drastically the resiliency of the data access layer and has the advantage to be without compromise or impact on the performances.
- The **"Multi-stream" strategy**. In this case, *libdavix* will first try to obtain the Metalink of the resource and then proceed to a multi-source parallel download of each referenced chunk of data from a different replica. This approach has the advantage to maximize the network bandwidth usage on the client side and to offer the same resiliency improvement than the fail-over strategy. However, it has for main drawback to overload considerably the servers.

The combined usage of *libdavix* for data analysis with a Replica catalogue or federation system supporting able to provide Metalink files like the DynaFed system (Dynamic Storage Federation) [19] enforces the global resilience of the I/O layer of any HPC application in a transparent manner. It provides the guarantee that a read operation on a resource will succeed as long as one replica of this resource is remotely accessible and referenced by the corresponding Metalink.

## 3   Performance Analysis

For our performance analysis, we executed a High energy analysis job based on ROOT framework [4] reading a fraction or the totality of around 12000 particles events from a 700 MBytes root file. This tests has been executed using both the XRootD framework and our libdavix solution as I/O layer.

Each test execution has been executed on WLCG through the Hammerloud Grid [40] performance testing framework.

The execution of the test is always done on a standard Worder Node configuration of WLCG.

For both XRootD and davix, each test is run against the same server instance with the following configuration: Disk Pool Manager(DPM) Storage system, 4 Core Intel Xeon CPU, 12 GBytes of RAM, Scientific Linux CERN 6, 1 GB/s network link.

Two test executions are separated by 20 min.

Our statistics are based on an average of 576 tests executions over a period of 12 days.

We compare here the global execution time of the analysis job over different network configurations:

- **LAN**: accessing the file over a gigabit Ethernet with low latency (latency <5 ms)
- **PAN-European network**: accessing the file over GEANT [24] between Switzerland and UK (latency <50 ms)

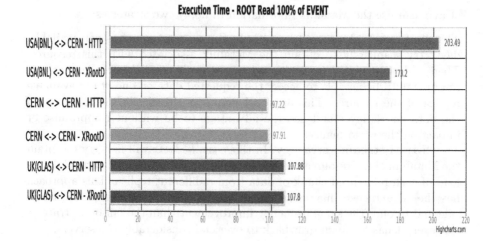

**Fig. 4.** Execution time of a ROOT data analysis job (less is better).

- **WAN**: accessing the file over Internet between Switzerland and USA (latency <300 ms)

In case of "CERN-CERN" data transfer, the (Fig. 4) shows that *libdavix* is respectively 0.7 % faster than XRootD in case of local access with high speed link and low latency. This shows that HTTP with *libdavix* can compete with a HPC specific I/O protocol on local area network and offers similar performances in term of data access time and data transfer rate.

In case of "UK(GLAS)-CERN", XRootd and *libdavix* offers sensibly the same performance.

In case of "USA(BNL)-CERN" data transfer, our tests shows that XRootD is in average 17.5 % faster than *libdavix* on Wide Area network links with high latency. This difference of performance comes mainly from the sliding windows buffering algorithm of XRootD which allows to minimize the number of network round trips executed. Network round trips are naturally extremely costly on high latency networks.

In a classical High Energy Physics grid computing model, a job is always sent close to the data that it will process. Data access are in this case made over a LAN with high speed and low latency.

In such model, these results are particularly encouraging for *libdavix* and HTTP I/O.

## 4    Conclusion

Our results shows that for a HPC I/O use case, our solution, *libdavix* can provide similar performance over low latency link to a HPC specific protocol like XRootD.

The usage of the HTTP multi-range feature allows to reduce drastically the number of parallel I/O operations and allows HTTP to compete with the aggressive caching strategy of the HPC specific protocols in case remote partial I/O operations on large data sets.

The lack of multi-plexing support in HTTP can be compensated by a session recycling system for HEP uses cases and allows to be retro-compatible with the existing network and service infrastructure.

Finally, the association of the Metalink with HTTP gives new possibilities for transparent error recovery in HPC data access and offers an interesting alternative to classical hierarchical data federations.

We have demonstrated in this paper with *libdavix* that an optimized I/O layer based on the HTTP protocol can be considered as a serious and performant alternative to the common HPC specific I/O protocols.

**Acknowledgments.** We thank Oliver Keeble, Martin Hellmich, Ivan Calvet and Alejandro Alvarez Ayllon for their contributions, commitment and testing dedicated the davix toolkit. Thank you also to Tigran Mkrtchyan for his contributions to TDavix-File. We also thank the ROOT development team for their support and help during the integration of davix to the ROOT analysis framework. Thank you to Olivier Perrin for his support and advises in this work.

# References

1. Davix project. doi:10.5281/zenodo.10761
2. Riak: an open source scalable data store, 5 June 2014
3. Anderson, J.C., Lehnardt, J., Slater, N.: CouchDB the Definitive Guide. O'Reilly Media Inc., Sebastopol (2010)
4. Antcheva, I., Ballintijn, M., Bellenot, B., Biskup, M., Brun, R., Buncic, N., Canal, P., Casadei, D., Couet, O., Fine, V., Franco, L., Ganis, G., Gheata, A., Maline, D.G., Goto, M., Iwaszkiewicz, J., Kreshuk, A., Segura, D.M., Maunder, R., Moneta, L., Naumann, A.: Root - a c++ framework for petabyte data storage, statistical analysis and visualization. Comput. Phys. Commun. **182**(6), 1384–1385 (2011)
5. Baset, S.A.: Open source cloud technologies. In: Proceedings of the Third ACM Symposium on Cloud Computing, SoCC '12, pp. 28:1–28:2. ACM, New York (2012)
6. Battle, R., Benson, E.: Bridging the semantic web and web 2.0 with representational state transfer (rest). Web Semant. **6**(1), 61–69 (2008)
7. Belshe, M., Peon, R.: Spdy protocol (2012)
8. Berners-Lee, T., Cailliau, R.: Worldwideweb: proposal for a hypertext project (1990). Accessed 26 February 2008
9. Berners-Lee, T., Fielding, R., Frystyk, H.: Hypertext transfer protocol-http/1.0 (1996)
10. Box, D., Ehnebuske, D., Kakivaya, G., Layman, A., Mendelsohn, N., Nielsen, H.F., Thatte, S., Winer, D.: Simple object access protocol (soap) 1.1 (2000)
11. Bresnahan, J., Link, M., Khanna, G., Imani, Z., Kettimuthu, R., Foster, I.: Globus gridftp: what's new in 2007. In: Proceedings of the First International Conference on Networks for Grid Applications, GridNets '07, ICST, Brussels, Belgium, Belgium. ICST (Institute for Computer Sciences, Social-Informatics and Telecommunications Engineering), pp. 19:1–19:5 (2007)

12. Bryan, A., Tsujikawa, T., McNab, N., Poeml, P.: The metalink download description format. Technical report, RFC 5854, June 2010
13. Buyya, R., Broberg, J., Goscinski, A.: Cloud Computing: Principles and Paradigms. Wiley Series on Parallel and Distributed Computing. Wiley, New York (2010)
14. Coarfa, C., Druschel, P., Wallach, D.S.: Performance analysis of tls web servers. ACM Trans. Comput. Syst. **24**(1), 39–69 (2006)
15. Daboo, C., Desruisseaux, B., Dusseault, L.: Calendaring extensions to webdav (caldav). Technical report, RFC 4791, March 2007
16. Dorigo, A., Elmer, P., Furano, F., Hanushevsky, A.: Xrootd-a highly scalable architecture for data access. WSEAS Trans. Comput. **1**(4.3) (2005)
17. Fielding, R.T.: Architectural styles and the design of network-based software architectures. Ph.D. thesis. AAI9980887 (2000)
18. Foster, I., Kesselman, C.: The globus toolkit. In: The Grid: Blueprint for a New Computing Infrastructure, pp. 259–278. Morgan Kaufmann Publishers, San Francisco (1999)
19. Furano, F., da Rocha, R.B., Devresse, A., Keeble, O., Ayllón, A.Á., Fuhrmann, P.: Dynamic federations: storage aggregation using open tools and protocols. J. Phys. Conf. Ser. **396**, 032–042 (2012). IOP Publishing
20. Gettys, J., Nielsen, H.F.: Smux protocol specification. Work In Progress (W3C Working Draft WD-mux-19980710) (1998)
21. Goland, Y., Whitehead, E., Faizi, A., Carter, S., Jensen, D.: Http extensions for distributed authoring-webdav (1999)
22. Heidemann, J.: Performance interactions between p-http and tcp implementations. SIGCOMM Comput. Commun. Rev. **27**(2), 65–73 (1997)
23. Worldwide LHC computing grid website. http://wlcg.web.cern.ch/
24. Geant project website. http://www.geant.net/
25. Metalink project website. http://www.metalinker.org/
26. A. Inc. Amazon s3 rest (2014)
27. A. Inc. Amazon ss (2014)
28. G. Inc. Spdy frequently asked questions, doesn't http pipelining already solve the latency problem? (2014)
29. Jackson, K.: OpenStack Cloud Computing Cookbook. Packt Publishing Ltd, Birmingham (2012)
30. Janssen, B.: Http-ng overview problem statement, requirements, and solution outline
31. Manyika, J., Chui, M., Brown, B., Bughin, J., Dobbs, R., Roxburgh, C., Byers, A.H.: Big data: the next frontier for innovation, competition, and productivity (2011)
32. Martin, J.: Managing the Data Base Environment, 1st edn. Prentice Hall PTR, Upper Saddle River (1983)
33. Natarajan, P., Iyengar, J.R., Amer, P.D., Stewart, R.: SCTP: an innovative transport layer protocol for the web. In: Proceedings of the 15th International Conference on World Wide Web, WWW '06, pp. 615–624. ACM, New York (2006)
34. H. RFC. Hypertext transport protocol RFC (1999)
35. Schmuck, F., Haskin, R.: GPFS: a shared-disk file system for large computing clusters. In: Proceedings of the 1st USENIX Conference on File and Storage Technologies, FAST '02. USENIX Association, Berkeley (2002)
36. Snia. CDMI specification (2014)
37. Spero, S.E.: Analysis of http performance problems, July 1994

38. Team, D.: Davix project website. http://dmc.web.cern.ch/projects/davix/home
39. Thakur, R., Gropp, W., Lusk, E.: Data sieving and collective i/o in romio. In: 1999 The Seventh Symposium on the Frontiers of Massively Parallel Computation, Frontiers '99, pp. 182–189, February 1999
40. van der Ster, D.C., Elmsheuser, J., Garcia, M.U., Paladin, M.: Hammercloud: a stress testing system for distributed analysis. J. Phys. Conf. Ser. **331**, 072036 (2011). IOP Publishing
41. Vukotic, I., Collaboration, A., et al.: Optimization and performance measurements of root-based data formats in the atlas experiment. J. Phys. Conf. Ser. **331**, 032032 (2011). IOP Publishing
42. White, T.: Hadoop: The Definitive Guide. O'Reilly Media, Sebastopol (2009)

# Topical Section Headings:
# Emerging Hardware

# Exploring Opportunities for Non-volatile Memories in Big Data Applications

Wei Wei[1,2(✉)], Dejun Jiang[1], Jin Xiong[1], and Mingyu Chen[1]

[1] Institute of Computing Technology, Chinese Academy of Sciences, Beijing, China
[2] University of Chinese Academy of Sciences, Beijing, China
weiwei01@ict.ac.cn

**Abstract.** Large-capacity memory system allows big data applications to load as much data as possible for in-memory processing, which improves application performance. However, DRAM faces both scalability and energy challenges due to its inherent charging mechanism. Thus, DRAM-based memory system incurs excessive cost to meet both capacity and energy requirements for the emerging big data workloads. Fortunately, non-volatile memories(NVMs) are emerging with the advanced features of better scalability and lower power leakage. Integrating NVMs into main memory is non-trivial as NVMs have a few weakness, such as asymmetric read and write latency and power. Designing memory system comprising both DRAM and NVMs requires to understand the memory access behaviors of big data applications. In this paper, we first investigate the memory access patterns of both typical big data workloads and traditional parallel workloads. By doing so, we show the read/write intensity as well as temporal/spatial locality of big data workloads. We then replay memory access traces of big data applications to DRAM simulator and PCM simulator, respectively. We explore design implications of hybrid memory comprising DRAM and PCM.

## 1 Introduction

Emerging big data applications need to process high volume of data. For instance, Facebook keeps 75 % of its non-image data in main memory [1]. Loading as much data into memory as possible avoids time-consuming disk IO operations, and thus accelerates application performance. Large-capacity memory system is expected to run big data applications. However, DRAM-based memory system faces both scalability and energy challenges due to its working mechanism. It is considered to be hard for DRAM to scale down to 20 nm [21]. Furthermore, DRAM consumes 20 % to 40 % of the total server energy in data center [12]. As such, DRAM-based memory system incurs excessive cost to provide large capacity to run big data applications. Fortunately, non-volatile memories, such as Phase Change Memory [14], Resistive RAM [8], and STT-RAM [6], are emerging to become alternative candidates to DRAM-based memory system. Compared to DRAM, NVMs have higher density without energy leakage. A few NVMs have comparable operation performance with DRAM. Therefore, NVMs are desirable to serve as large-capacity memory system for big data applications.

© Springer International Publishing Switzerland 2014
J. Zhan et al. (Eds.): BPOE 2014, LNCS 8807, pp. 209–220, 2014.
DOI: 10.1007/978-3-319-13021-7_16

However, NVMs also have a few drawbacks. For example, PCM has asymmetric read/write energy consumption. The write latency of PCM is higher than DRAM. When adopting NVMs in memory system for running big data applications, one needs to understand the memory access patterns of big data applications. Memory access patterns, such as staged bandwidth, read/write ratio, temporal and spatial locality, are able to provide design implications to NVM-based memory system. Without understanding these access patterns, one may allocate PCM spaces for write-intensive data objects, which results in increased memory performance due to the relatively slow PCM writes. Although there are some studies characterizing the big data workloads, the opportunities of NVM in big data area have not been examined as far as we know.

In this paper, we explore the opportunities of NVMs in big data applications. We conduct a set of experiments to study the detailed memory access patterns of typical big data applications. By gathering memory traces of big data applications, we analyze the staged bandwidth, read/write ratio, temporal locality and spatial locality. In addition, we develop a PCM simulator based on DRAM-Sim2 [16] and replay the memory access traces to PCM simulator. By doing so, we provide design implications for integrating PCM into memory system. The contribution of this paper is as follow.

1. We conduct experiments to analyze the memory access patterns of both big data workloads and traditional parallel workloads. We observe that big data workloads usually have similar read intensity and write intensity. In addition, we observe that big data workloads have weaker temporal and spatial locality compared to traditional ones.
2. We use PCM simulator to explore the opportunities of NVMs, especially PCM, to serve as memory system for running big data applications. We show the great opportunities for NVM to reduce energy consumption in big data applications.
3. We propose the adoption of DRAM-PCM hybrid memory as a desirable design choice when integrating NVMs into main memory. We argue that this design choice can fully utilized advantages of PCM while reduce the negative impact of its long write latency. To make the design more practicable, we suggest that carefully designed data placement are essential.

The rest of the paper is organized as follows. Section 2 presents related works. In Sect. 3, we describe the applications and methods used in our experiment evaluations. Section 4 shows the memory access pattern analysis results as well as the simulation results in both PCM and DRAM memory. We conclude the paper in Sect. 5.

## 2   Related Work

### 2.1   Using NVMs in Computer Memory Architecture

Compared to DRAM, NVMs have higher density. Meanwhile, NVMs are promising to have comparable read/write performance as DRAM, and even SRAM.

Therefore, a few research works integrate NVMs into cache hierarchy or main memory to increase the capacity and reduce energy cost. However, current NVMs also have several weaknesses, such as high dynamic energy and high write latency. Smullen et al. [17] addresses the issue by redesigning the MTJ to use a smaller free layer. Li et al. [11] adopt STT-RAM together with SRAM to construct a hybrid adaptive on-chip cache to provide low power consumption and low access latency. Recent works also focus on combining a small amount of DRAM and a large amount of PCM to provide a low latency and energy memory system [5,9,15]. These works mainly focus on the optimization technologies of PCM to be adopted in main memory. These works target traditional workloads. However, this paper mainly focuses on the opportunities of NVMs in big data applications.

## 2.2  The Role of NVMs

Similar to our work, recently, a few works also explore the role of NVMs in different types of workloads. Li et al. [10] perform a detailed analysis to the access patterns of memory objects in stack, heap and global data for real large-scale scientific applications. Their results reveal a lot of opportunities for using NVRAM in scientific applications. Caulfield et al. [3] explores several options for connecting solid-state storage to the host system to evaluate the role of NVMs in storage level for high-performance and IO-intensive computing. Essen et al. [18] develops a NVM simulator to model the impact of future generations of I/O-bus-attached NVM on HPC application performance at scale. This paper differs from these works in that we mainly evaluate the performance and energy improvement when one adopts NVMs in main memory for big data workloads.

## 2.3  Analysis of Big Data Workloads

Recently, a few research efforts have been made to characterize and understand the features of big data workloads. Chang et al. [4] examine the implications of big data workloads on system design. Jia et al. [7] characterize micro-architectural characteristics of big data workloads on the systems equipped with modern superscalar out-of-order processors. Dimitrov et al. [13] profile the memory access patterns of big data workloads, such as memory footprint, CPI, and cache misses etc. However, these works do not examine the opportunities of NVMs (e.g. PCM) on big data workloads.

## 3  Applications and Methodology

In this section, we present the evaluated applications and the evaluation method.

**Table 1.** Features of Applications

| Application scenarios | Applications | Data type | Data size | Software stack | Application type | Memory footprint |
|---|---|---|---|---|---|---|
| Relational query | Join query | Table | 7.1 GB | Impala,Shark, MySQL,Hive | Realtime analytics | 7688 MB |
| Relational query | Select query | Table | 7.1 GB | Impala,Shark, MySQL,Hive | Realtime analytics | 7867 MB |
| Search engine | PageRank | Graph | 4.2 GB | Hadoop,MPI Spark | Offline analytics | 7892 MB |
| Social network | K-means | Graph | 4.1 GB | Hadoop,MPI Spark | Offline analytics | 7283 MB |
| Micro benchmarks | Terasort | Text | 2 GB | Hadoop MPI Spark | Offline analytics | 7215 MB |
| Traditional applications | Volrend | N/A | N/A | N/A | Parallel applications | 6289 MB |
| Traditional applications | Radiosity | N/A | N/A | N/A | Parallel applications | 7314 MB |

## 3.1 Applications

We select 5 workloads from the BigDataBench [19] as the big data workload set, which are representative in big data scenarios. Table 1 shows the features of these workloads. Two of the workloads are relational queries, which represents realtime analytics workload. The other three represent offline data analytics workload, which are PageRank, K-means, and Terasort. We also select two applications radiosity and volrend from SPLASH-2 benchmark [20] as the traditional parallel workload set. We conduct experiments using the two sets of workloads and compare their memory access patterns.

## 3.2 Methodology

We first gather memory traces of the 7 workloads to analyze their memory access patterns. For each workload, we measure stage bandwidth, read counts, and write counts. The stage bandwidth is measured every 10 million requests at runtime, which shows the bandwidth trend of each workload. We calculate the total bandwidth, read bandwidth and write bandwidth for each stage. We calculate the read-to-write ratio *rwRatio* for each page to show the write intensity of each workload. Since memory access pattern is related to CPU cache design, we further analyze the temporal locality of each workload by counting the access intervals of each page. We count the page references and the reference distribution for each workload to show its spatial locality.

We use a 5 nodes cluster to run the 4 Hadoop-based big data workloads, in which one is the master node and others are slave nodes. Each node is configured with 4 Intel Xeon E5 processors, 8 GB of DDR3 memory, and 1TB Disk. The big data workload K-means runs using one single node within the cluster as it does not employ the Hadoop stack. Similarly, the two traditional workloads run on one node within the cluster. We use a lightweight hardware tool HMTT [2] to gather memory traces. The memory trace for a workload includes the timestamp of a memory access, the memory accessing address, and the accessing size. We develop tools to derive evaluation metrics from the raw data.

In order to explore the opportunities of NVM in big data workloads, we use the collected memory traces to evaluate the application performance and energy consumption when adopting PCM as memory system. We develop a PCM simulator PCMSim based on DRAMSim2 [16] to simulate the results when running workloads on PCM-based memory. The PCM simulator receives the collected memory traces as input and outputs the statistical results, such as access latency and energy. To compare the performance and energy results with DRAM-based memory system, we also replay memory traces to the DRAMSim2 simulator. Table 2 shows the configurations of the two simulators.

**Table 2.** Features of DRAMSim2 and PCMSim

| Features | | Values | |
|---|---|---|---|
| | | DRAM | PCM |
| Device | Device width | 16 | 16 |
| | Rank size | 512(MB) | 512(MB) |
| Latency | Burst latency | 9.33(NS) | 6.47(NS) |
| | Read latency | 20(NS) | 36.28(NS) |
| | Write latency | 20(NS) | 210.54(NS) |
| Energy | Burst read energy | 23.55(PJ) | 7.36(PJ) |
| | Burst write energy | 23.55(PJ) | 7.36(PJ) |
| | Array read energy | 23.76(PJ) | 20.86(PJ) |
| | Array write energy | 28.28(PJ) | 20(PJ) |
| Others | Rowbuffer policy | close | close |
| | Transaction queue | 32 | 32 |
| | Command queue | 32 | 32 |

# 4   Experimental Results

In this section, we first show the analysis results of memory access patterns of both big data workloads and traditional parallel workloads. Then, we present the simulation results by comparing running workloads on PCM memory and DRAM memory separately.

## 4.1   Memory Access Patterns

**Stage Bandwidth.** Figure 1 shows the stage bandwidths of 6 workloads. The bandwidth result of workload *join query* is similar to that of workload *select query*.

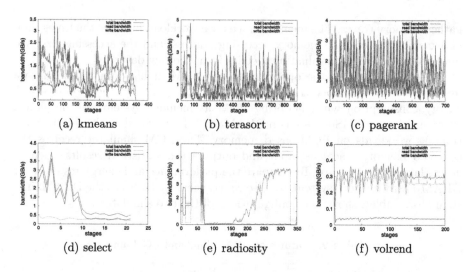

**Fig. 1.** Stage bandwidth of workloads

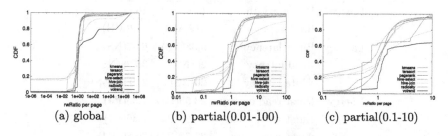

**Fig. 2.** CDF of rwRatio among pages

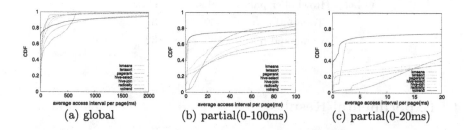

**Fig. 3.** CDF of average access interval

Thus, we only show the result of workload *join query*. For workload *join query*, its write bandwidth is higher than the read bandwidth in the first half stages. Its write bandwidth then drops and becomes close to the read bandwidth in the second half stages. On average, all other workloads have higher read bandwidth than write bandwidth. Especially, Fig. 1(e) shows that the traditional workload *radiosity* become read-dominant in the second half stages. Meanwhile, Fig. 1(f) shows that workload *volrend* is read-intensive in the whole running process. We further calculate the ratio of maximum read bandwidth to maximum write bandwidth for each workload. The ratio results show that big data workloads have similar counts of read and write operations. For example, the ratio results of workloads *Kmeans, Terasort,* and *Pagerank* are 3.27, 32.41, and 3.12 respectively. In contrast, the ratio result of workload *radiosity* is 135.25. Therefore, traditional parallel workloads can exhibit read-intensive feature, while big data workloads usually exhibit similar read intensity and write intensity.

**rwRatio.** In addition to calculate the ratio of read bandwidth to write bandwidth, we further the rwRatio of each page for all workloads. By doing so, we can investigate the read/write features of these workloads at finer granularity. We plot the CDF curves of rwRatios for all workloads in Fig. 2(a). We observe that the rwRatio of each page is between 0.01 and 100 in almost all workloads. As such, we show the enlarged portion of rwRatio just between 0.01 and 100 in Fig. 2(b). Furthermore, we show the enlarged portion of rwRatio just between 0.1 and 10 in Fig. 2(c).

In traditional workloads, *volrend* has 60 % pages whose rwRatios are between 1 and 10, and 35 % pages whose rwRatios are greater than 100. Therefore, it is desirable to allocate pages in PCM for workload *volrend* to achieve near-DRAM read performance while keeping low energy consumption. However, the rwRatio results of big data workloads are much more diverse. For example, workload *join query* has a very wide distribution of rwRatio. The pages whose rwRatios are greater than 10 and smaller than 0.1 occupy 20 % respectively. Therefore, these pages are expected to be allocated in PCM and DRAM separately. The rwRatios of other big data workloads are nearly between 0.1 and 10, which indicates that most pages have equal read/write references. For example, workloads *Pagerank* and *Kmeans* both have nearly 75 % pages whose rwRatios are between 1 and 10.

**Temporal and Spatial Locality.** Figure 3(a) plots the CDF curves of average access interval for pages in all workloads. In most workloads except *kmeans*, 80 % pages have the average access interval below 500 ms. In order to make more detailed analysis, we plot the enlarged distributions of average access interval between 0 and 100 ms in Fig. 3(b) and that between 0 and 20 ms in Fig. 3(c). For workload *radiosity* and *volrend*, 60 % pages have the average access interval below 2.5 ms. In contrast, there are no more than 10 % pages whose average access intervals are below 5 ms in *kmeans* and *pagerank*. Similarly, workloads *terasort, select query,* and *join query* have about 20 % pages whose average access intervals are below 5 ms. As such, compared to traditional workloads, big data workloads have weaker temporal locality which requires changes in CPU LLC cache algorithm design.

In order to analyze the spatial locality of these workloads, we calculate the references of each page in all workloads. We sort these pages according to their references. We then calculate the percentage of the pages that contribute 80 % of total accesses for each workload. We show the results in Fig. 4. Less than 10 % of

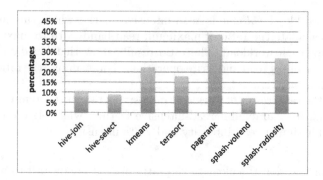

**Fig. 4.** The percentages of the pages contributing 80 % references

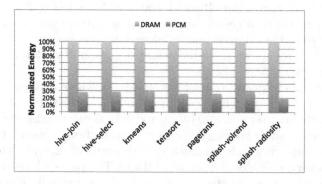

**Fig. 5.** Comparison of Energy consumption

pages contribute 80 % references for workloads *hive-join* and *hive-select*, which exhibits strong spatial locality. Hybrid memory comprising DRAM and NVM either employs NVM as the extension of DRAM or places DRAM in front of NVM as a cache. As such, the hybrid memory architecture employing DRAM as a cache is more appropriate for the *Realtime Analysis* workloads, such as *hive-join* and *hive-select*. For workloads *pagerank*, the 80 % references are contributed by more than 35 % pages, which exhibits weak spatial locality. Considering the high volume processed data processed, the big data workload with weak spatial locality expects carefully-designed cache policy to avoid degraded performance due to disk IO overheads.

## 4.2   PCM Simulation Results

In this section, we explore the opportunities of NVMs as main memory to run big data applications. We use PCM as the candidate memory device. We show the energy and average latency results when directly running big data workloads in both DRAM and PCM memory systems. We use the conventional memory management stack when using PCM memory. We use DRAM memory as the baseline and normalize all results in PCM to the baseline.

**Energy and Latency.** Figure 5 shows the energy results of all workloads running in the two types of memory devices. Since PCM has no refresh energy consumption, all workloads achieve lower energy cost in PCM. In big data workloads, the maximum energy reduction is 74.28 % for *terasort*, meanwhile the minimum energy reduction also reaches 70.98 % for *kmeans*. Big data applications typically require large-capacity memory to accelerate data processing. By adopting low-energy NVMs into memory, one can not only fully utilize the high density of NVM chips to increase capacity, but also significantly reduce the energy cost of memory system and thus the total cost of data-center servers.

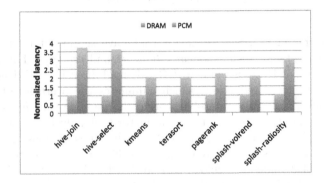

**Fig. 6.** Average latency comparison in DRAM and PCM

However, as shown in Fig. 6, the average latency observed in PCM memory is higher than that in DRAM memory for all workloads. The minimum average latency increase in PCM is 1 times higher than that in DRAM for workload *kmeans*. Especially, the average latencies of *hive-join* and *hive-select* increase by up to 3.6 times, compared to the baseline. This is because both workloads have a large fraction of write operations (as shown in Sect. 4.1), and the long write latency of PCM correspondingly results in the increase of their average latencies. Although PCM provides better scalability and lower energy to memory system, the long write latency limits its application in real-world systems. However, note that we employ the conventional memory management to the underlying PCM memory. Moreover, we only evaluate the performance with the memory model comprising only PCM. Actually, in order to avoid the negative impact of long write latency of PCM, one can use hybrid memory architecture to exploit advantages of both DRAM and PCM while reduce the impacts of their disadvantages. Correspondingly, one should carefully design data placement policies when using hybrid memory, instead of simply applying current memory management stack.

**ED and ED2.** We further plot the *energy-delay* and $energy-delay^2$ results for all workloads in DRAM and PCM. The two metrics is able to show the trade-off between performance gain and energy reduction when designing an efficient memory system. Figure 7 shows the values of ED and ED2 in the two memory

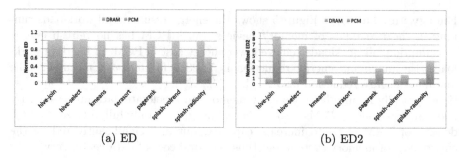

(a) ED                                        (b) ED2

**Fig. 7.** Comparison of ED and ED2

devices. In terms of ED, the workloads that have significant read operations perform better in PCM, such as *kmeans*, *terasort*, and *pagerank*. This is because the read latency of PCM is close to DRAM, read-intensive workloads benefit more from the energy reduction, which eliminates the overhead of increased latency. However, for write-intensive workloads *hive-join* and *hive-select*, the increased write latency becomes the main factor that reduces the memory system efficiency. Figure 7(b) shows the ED2 results. It is obvious that the long write latency of PCM significantly reduces the efficiency of memory system. For example, the ED2 value of workload *hive-join* in PCM is 8.3 times higher than that in DRAM. For latency-sensitive workloads, it is not a good design choice to completely replace DRAM memory with PCM memory though PCM can provide larger capacity. We suggest that one should consider hybrid memory system to allow more practical usage of PCM.

## 5   Conclusion

This paper examines the opportunity of NVMs in big data workloads, which is not explored in previous research. We perform memory access pattern analysis towards both emerging big data workloads and traditional parallel workloads. We investigate the read-to-write ratios as well as temporal and spatial locality of both workload sets. We show that big data workloads exhibit weak temporal and spatial locality compared to traditional workloads. In order to explore the opportunities of NVMs as memory system to run big data workloads, we further compare the performance and energy results by directly running big data workloads in DRAM memory and PCM memory. We evaluate the memory efficiency by replaying memory access traces to both DRAM and PCM simulators, respectively. By doing so, we observe the great opportunity for NVMs to significantly reduce the energy consumption for big data workloads. We also propose hybrid memory architecture as a desirable design choice for integrating NVMs into main memory system. This design choice can exploit the advantages of NVMs, and meanwhile avoid their unpleasant features.

# References

1. International technology roadmap for semiconductors emerging research devices (2011). http://www.itrs.net
2. Bao, Y., Chen, M., Ruan, Y., Liu, L., Fan, J., Yuan, Q., Song, B., Xu, J.: HMTT: a platform independent full-system memory trace monitoring system. In: Proceedings of the International Conference on Measurement and Modeling of Computer Systems, pp. 229–240 (2008)
3. Caulfield, A.M., Coburn, J., Mollov, T., De, A., Akel, A., He, J., Jagatheesan, A., Gupta, R.K., Snavely, A., Swanson, S.: Understanding the impact of emerging non-volatile memories on high-performance, IO-intensive computing. In: Proceedings of the International Conference for High Performance Computing, Networking, Storage and Analysis, pp. 1–11 (2010)
4. Chang, J., Lim, K.T., Byrne, J., Ramirez, L., Ranganathan, P.: Workload diversity and dynamics in big data analytics: implications to system designers. In: Proceedings of the 2nd Workshop on Architectures and Systems for Big Data, pp. 21–26 (2012)
5. Dhiman, G., Ayoub, R., Rosing, T.: PDRAM a hybrid PRAM and DRAM main memory system. In: Proceedings of the Design Automation Conference, pp. 664–669 (2009)
6. Dieny, B., Sousa, R., Bandiera, S., Castro Souza, M., Auffret, S., Rodmacq, B., Nozieres, J., Herault, J., Gapihan, E., Prejbeanu, I., et al.: Extended scalability and functionalities of MRAM based on thermally assisted writing. In: Proceedings of the International Electron Devices Meeting, pp. 1–3 (2011)
7. Jia, Z., Wang, L., Zhan, J., Zhang, L., Luo, C.: Characterizing data analysis workloads in data centers. CoRR 1307.8013 (2013)
8. Kawahara, A., Azuma, R., Ikeda, Y., Kawai, K., Katoh, Y., Tanabe, K., Nakamura, T., Sumimoto, Y., Yamada, N., Nakai, N., Sakamoto, S., Hayakawa, Y., Tsuji, K., Yoneda, S., Himeno, A., Origasa, K., Shimakawa, K., Takagi, T., Mikawa, T., Aono, K.: An 8mb multi-layered cross-pointReRAM macro with 443MB/s write throughput. In: Proceedings International Solid-State Circuits Conference, pp. 432–434 (2012)
9. Lee, H.G., Baek, S., Nicopoulos, C., Kim, J.: An energy- and performance-aware DRAM cache architecture for hybrid DRAM/PCM main memory systems. In: Proceedings of the International Computer Design Conference, pp. 381–387 (2011)
10. Li, D., Vetter, J.S., Marin, G., McCurdy, C., Cira, C., Liu, Z., Yu, W.: Identifying opportunities for byte-addressable non-volatile memory in extreme-scale scientific applications. In: Proceedings International Parallel and Distributed Processing Symposium, pp. 945–956 (2012)
11. Li, J., Xue, C., Xu, Y.: STT-RAM based energy-efficiency hybrid cache for CMPs. In: Proceedings of the International Conference on VLSI and System-on-Chip, pp. 31–36 (2011)
12. Lim, K., Ranganathan, P., Chang, J., Patel, C., Mudge, T.N., Reinhardt, S.: Understanding and designing new server architectures for emerging warehouse-computing environments. In: Proceedings of Annual International Symposium on Computer Architecture, pp. 315–326 (2008)
13. Martin Dimitrov, Karthik Kumar, P.L., Viswanathan, V.: Memory system characterization of big data workloads. In: The 1st Workshop on Benchmarks, Performance Optimization, and Emerging hardware of Big Data Systems and Applications (2013)

14. Nirschl, T., Phipp, J.B., Happ, T.D., Burr, G.W., Rajendran, B., Lee, M.H., Schrott, A., Yang, M., Breitwisch, M., Chen, C.F., Joseph, E., Lamorey, M., Cheek, R., Chen, S.H., Zaidi, S., Raoux, S., Chen, Y.C., Zhu, Y., Bergmann, R., Lung, H.L., Lam, C.: Write strategies for 2 and 4-bit multi-level phase-change memory. In: Proceedings of the International Electron Devices Meeting, pp. 461–464 (2007)

15. Ramos, L.E., Gorbatov, E., Bianchini, R.: Page placement in hybrid memory systems. In: Proceedings of the International Conference on Supercomputing, pp. 85–95 (2011)

16. Rosenfeld, P., Cooper-Balis, E., Jacob, B.: DRAMSim2: a cycle accurate memory system simulator. Comput. Archit. Lett. **10**, 16–19 (2011)

17. Smullen, C., Mohan, V., Nigam, A., Gurumurthi, S., Stan, M.: Relaxing nonvolatility for fast and energy-efficient stt-ram caches. In: Proceedings of the International Symposium on High Performance Computer Architecture, pp. 50–61 (2011)

18. Van Essen, B., Pearce, R., Ames, S., Gokhale, M.: On the role of NVRAM in dataintensive architectures: an evaluation. In: Proceedings of the International Parallel Distributed Processing Symposium, pp. 703–714 (2012)

19. Wang, L., Luo, C., He, Y., Zhan, J., Zhan, K., Li, X., Zhu, Y., Zhang, S., Yang, Q., Qiu, B., Jia, Z.: Bigdatabench: a big data benchmark suite from internet services. In: Proceedings of the International Symposium On High Performance Computer Architecture (2014)

20. Woo, S.C., Ohara, M., Torrie, E., Singh, J.P., Gupta, A.: The SPLASH-2 programs: characterization and methodological considerations. In: Proceedings of the International Symposium on Computer Architecture, pp. 24–36 (1995)

21. Xue, C.J., Zhang, Y., Chen, Y., Sun, G., Yang, J.J., Li, H.: Emerging non-volatile memories: opportunities and challenges. In: Proc. International Conference on Hardware/Software Codesign and System Synthesis, pp. 325–334 (2011)

# Author Index

Printed in the United States
By Bookmasters

Printed in the United States
By Bookmasters